Made in America

Made in America

Science, Technology, and American Modernist Poets

LISA M. STEINMAN

Yale University Press
New Haven and London

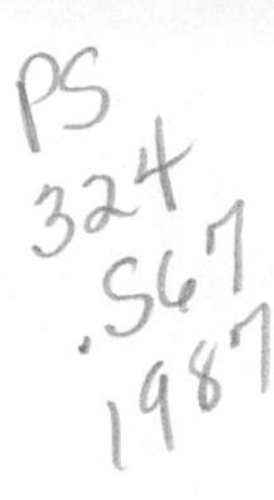

Published with assistance from the Louis Stern Memorial Fund

Designed by James J. Johnson and set in Electra types by The Composing Room of Michigan.
Printed in the United States of America by Halliday Lithograph, West Hanover, Mass.

Library of Congress Cataloging-in-Publication Data

Steinman, Lisa Malinowski, 1950–
Made in America.

Bibliography: p.
Includes index.
1. American poetry—20th century—History and criticism. 2. Literature and science. 3. Literature and technology. I. Title.
PS324.S67 1987 811′.5′09356 86–26550
ISBN 0–300–03810–0 (alk. paper)

The paper in this book meets the guidelines for permanence and durability of the Committee on Production Guidelines for Book Longevity of the Council on Library Resources.

10 9 8 7 6 5 4 3 2 1

■ ■ ■ ■ ■ ■ ■ ■ ■ ■ ■ ■ ■ ■ ■ ■ ■ ■ ■

New York! New York! I should like to inhabit you!
I see there science married
To industry,
In an audacious modernity
—Arthur Cravan, "Sifflet," *The Soil* 1.36

Contents

Acknowledgments

I have always enjoyed reading acknowledgments, even recognizing their formulaic nature. In the process of writing this book, I have come to understand first hand what those formulas reveal, namely that scholarly writing is not a solitary individual effort, but draws on the time, resources, and ideas of a large community of people. For example, the staff members and librarians at all the library collections in which I worked were more than generous with their time and willingness to make suggestions. I would especially like to thank Robert Bertholf, curator of the SUNY Buffalo Poetry / Rare Books Collection, as well as Marilyn Kierstead and Sam Sayre of the Reed College Library. My discussions and correspondence about Marianne Moore with Pat Willis, curator of the Marianne Moore Collection at the Rosenbach Museum & Library, were invaluable.

I cannot possibly mention the names of all the friends and colleagues who have inspired, helped, or encouraged me at various stages in the production of this book; however, I would like specifically to thank a few of those whose conversations with me I found particularly important: Bonnie Costello, Joan Richardson, Stephen Tapscott, Rich Lowry, and Casey Blake. I owe a special debt also to Theodora Graham, who suggested that I give a paper on Williams and science in

1977, and who offered much useful and detailed criticism as I was completing this manuscript. Reeve Parker, Neil Hertz, Alan Trachtenberg, and George Lensing, at different times, have gone out of their way to encourage me in this study, as has Ellen Graham, who has been an ideal editor. Finally, all of the Reed College students who have asked difficult questions, put up with me as I struggled to clarify my ideas, and in some cases taught me deserve thanks—especially the class that first asked me how Stevens and Williams felt about science, a question that then preoccupied me for almost ten years. In many ways, my largest debts are to William Ray, who read and argued with me through many drafts of this book, and to Jim Shugrue, without whose faith in me and in the importance of poetry I would not have written this book. All errors, needless to say, are my own.

I would also like to acknowledge the financial support I received from Howard and Jean Vollum, from Reed College, and from the National Endowment for the Humanities.

Portions of chapter 4 first appear in *The William Carlos Williams Review* in the fall of 1984, and they are reprinted here with permission from the *Williams Review*. A revised version of this same chapter appeared in the *Bucknell Review: Science and Literature*, edited by Harry R. Garvin (Lewisburg, Pennsylvania: Bucknell University Press, 1983), pp. 132–58; it is reprinted here by courtesy of the Associated University Presses. Chapter 5 first appeared in slightly different form in the fall / summer 1984 issue of *Twentieth Century Literature*, and appears here with permission. Finally, portions of chapter 6 first appeared in *The Wallace Stevens Journal*, 10(Spring 1986); they are reprinted with permission of that journal.

Materials from the Yale Collection of the William Carlos Williams papers and of the Alfred Stieglitz papers are reprinted with permission from the Collection of American Literature, the Beinecke Rare Book and Manuscript Library, Yale University. Materials from the Buffalo Poetry Collection are reproduced with the permission of the Poetry / Rare Books Collection, University Libraries, State University of New York at Buffalo. Previously unpublished material by William Carlos Williams is reproduced by permission of William Eric Williams and Paul H. Williams and by permission of New Directions Publishing Corporation, agents. Unpublished material by

Wallace Stevens is reproduced by permission of the Huntington Library, San Marino, California. New Directions has also granted permission for reprinting published material by William Carlos Williams and Ezra Pound. Unpublished material by Marianne Moore is printed here with permission from Clive E. Driver, Literary Executor of the Estate of Marianne C. Moore; the Marianne Moore materials from the Rosenbach Museum & Library are printed with permission from them. The letter from Lewis Mumford to Alfred Stieglitz is reproduced with the permission of the Alfred Stieglitz Collection, the Beinecke Rare Book and Manuscript Library, Yale University.

Abbreviations

Williams's work will be cited as follows:

A	*The Autobiography of William Carlos Williams* (1951; reprint ed. New York: New Directions, 1967).
CEP	*The Collected Earlier Poems* (1951; reprint ed. New York: New Directions, 1966).
CLP	*The Collected Later Poems* (rev. ed. 1963; reprint ed. New York: New Directions, 1967).
EK	*The Embodiment of Knowledge* (New York: New Directions, 1974).
I	*Imaginations* (New York: New Directions, 1970).
IAG	*In The American Grain* (1925; rev. ed. New York: New Directions, 1966).
SL	*The Selected Letters of William Carlos Williams*, ed. John C. Thirlwall (New York: McDowell, Obolensky, 1957).
SE	*Selected Essays of William Carlos Williams* (1954; reprint ed. New York: New Directions, 1969).
SSA	*Interviews with William Carlos Williams: "Speaking Straight Ahead,"* ed. Linda Wagner (New York: New Directions, 1976).

Pat — *Paterson* (New York: New Directions, 1963).
PB — *Pictures From Brueghel and Other Poems* (1962; reprint ed. New York: New Directions, 1967).
VP — A *Voyage to Pagany* (1928; reprint ed. New York: New Directions, 1970).

Moore's work will be cited as follows:

Coll — *Collected Poems* (New York: Macmillan, 1951).
Comp — *The Complete Poems of Marianne Moore* (New York: Macmillan, 1967).
Pred — *Predilections* (New York: Viking Press, 1955).
R — A *Marianne Moore Reader* (New York: Viking Press, 1961).
Sel — *Selected Poems of Marianne Moore* (New York: Macmillan, 1935).

Stevens's work will be cited as follows:

CP — *The Collected Poems of Wallace Stevens* (New York: Alfred A. Knopf, 1954).
L — *Letters of Wallace Stevens*, ed. Holly Stevens (New York: Alfred A. Knopf, 1966).
NA — *The Necessary Angel* (New York: Vintage, 1951).
OP — *Opus Posthumous*, ed. Samuel French Morse (New York: Alfred A. Knopf, 1957).
Palm — *The Palm at the End of the Mind*, ed. Holly Stevens (New York: Vintage Books, 1972).
SP — *Souvenirs and Prophecies: The Young Wallace Stevens*, ed. Holly Stevens (New York: Alfred A. Knopf, 1977).

Introduction

THE primary focus of this book is on modernist poetry and defenses of it written by William Carlos Williams, Marianne Moore, and Wallace Stevens between 1910 and 1945. I will say more about why I have chosen these three poets at the end of this introduction, but for all three, a defense of their art involved a defense of the place and value of such poetry in modern America.

I begin, therefore, with an examination of the American context in which Williams, Moore, and Stevens wrote. Scholars such as Meredith Neil, Alan Trachtenberg, and Jackson Lears have described the low esteem in which America traditionally held the arts and the way changes in the structure of American business and society after 1860 made the role of art and literature in the United States problematic.[1] By the late nineteenth century, American poetry was seen as effete and as divorced from the concerns of the work place and of everyday life. The first third of the twentieth century was filled with debates, discussed in detail in chapter 1, about how to improve American art and the quality of life in America. From this period, the writings of the so-called Young Intellectuals, men like Van Wyck Brooks and Randolph Bourne, were especially influential for Williams, Moore, and Stevens. This group identified American modernity as a central cause of the ill health of American life and letters, and one cure that they proposed was to remarry creative intelligence with the everyday reality of modern America.

The way American modernity was defined leads to a second topic of this study, namely how and why Williams, Moore, and Stevens were engaged in defining the relationship of poetry to science and technology. As I will argue, at least from the Romantic period, Anglo-American writers felt that science and philosophy had called poetry's importance into question, but in America defense of the arts was further complicated by an emphasis on practical concerns. By 1920, modernity was firmly linked with commercial and scientific development, in which America was thought to excel. The national respect for the *real world* of science and technology was sometimes linked with America's history as a relatively new country that needed to pay attention to practical concerns. Yet, this respect was also associated with the philosophical tradition of American pragmatism, closely affiliated with logical positivism, which emphasized science and factual truths over metaphysics or theory.[2]

Many disciplines, including poetry, tried to borrow the prestige of science and technology in order to declare a place for their own work in an American context. Some also argued that the arts, to be relevant, had to address the issues of the practical and technological world in which people lived. At the same time, most people saw themselves as defenders of art and human values against a *scientific age* (also called a *machine age*) and most agreed with Bertrand Russell that the problem of living humanely in the modern world could "only be solved by a community which used[d] machines without being enthusiastic about them."[3] Thus, there were contradictions in the positions taken by men such as Brooks or Dewey, who wanted both to use and to resist the effects of science and technology. Those who were most convincing in their descriptions of how poetry might participate in the everyday life of modern America were self-conscious about these contradictions.

Here and in the first two chapters, the terms *science* and *technology* are used interchangeably in part because the writings examined tend to lump the two together, at least until 1920 when Einsteinian physics changed the popular image of science. When science and technology are distinguished at all, science is usually characterized as a theoretical discipline, primarily concerned with understanding the world and without an eye to practical applications. Technology is seen as applied science or as the actual tools and products resulting from the application of scientific discoveries, with tools and products including not only machines but structures such as skyscrapers. Writers also invoke the ways in which technological hardware changed society—with assembly lines, for example—or even the inventions of systems such as those used to improve filing.[4] Science, then, can be separated from technology. More often, however, science and technology are used to refer indiscriminately to pure and applied science, to technological products of commercial value, and to those values and institutions associated with technological advances.

The identification of science with technology was not simple confusion or carelessness. Metaphorically, the world of Newtonian physics was easily equated with literal mechanisms. Historically, too, science and technology had in fact become interdependent. As Howard P. Segal explains in *Technological Utopianism in American Culture*, technology originally referred to the means of controlling the world (or *know-how*) while science referred to understanding (or *know-why*).[5] The *Oxford English Dictionary* indicates that technology did not narrowly refer to applied science until the 1880s and that date is revealing: the close connection between science and technology is a feature of the modern world, reflected in the change in the language.[6] The complexity of references to science between 1900 and 1920 also acknowledges the modernization of society and of scientific disciplines.

American writers often used the multiple meanings of the word science. They celebrated American scientific creativity and likened it to poetic understanding, even as they depended on the identification of science with technological know-how (which the public valued above theoretical science) to give force to their arguments. Yet many writers—even those who saw the value of the breakthroughs made possible by scientific discoveries—objected to the social values and institutions connected with technological progress. In other words, they defended poetry by comparing it to science and by appealing implicitly to values they did not endorse.

The situation of American modernist writers such as Williams, Stevens, and Moore was complicated by their ties to a movement in art and literature that took both visual images and a description of a style from the look of machinery and urban life. A serious examination of the international modernist movement is beyond the scope of this book, but a brief explanation of what modernism was understood to encompass is necessary. Its rhetoric, celebrations of motion, and uses of a machine aesthetic are particularly important to the argument of chapter 2. Many of the catchwords of modernism—efficiency, cleanliness, structure, mechanism, and motion—were perceived as being related to distinguishing features of industrial societies and landscapes. Moreover, the admiration of a machine aesthetic was a prominent feature of the modernist sensibility. And although it did not necessarily entail taking machines as subjects, the rhetoric—including the very name machine aesthetic—suggests the way this movement implicitly associated itself with social and historical modernity and especially with technological modernity, which in turn was identified with America.[7]

Modernism, then, promoted an aesthetic that was seen as uniquely American. Yet modernist work was not well received in the United States, and, conversely, American art and literature were perceived as inferior. So, American modernists, who were aware of their ties to a larger, international movement, faced a dilemma: How could they find an audience for the new in their work and also show that original work could be produced in America? In the first two decades of the century, the answer for many—including Ezra Pound, Marsden Hartley, Robert McAlmon, Charles DeMuth, and Mina Loy—was simply to leave for Europe.

Those who did not leave felt a pressing need to defend and define the distinctive qualities of American modernism. One aspect of the modernist style had much in common with what John Kouwenhoven calls the *vernacular tradition,* as exemplified in the clean lines of the Corliss engine and balloon frame houses.[8] Inventiveness in other areas, like engineering, was also related to creativity in the arts, especially by those whose style was identified with the products of American technology. With some justice, artists and writers could claim that their work sprang from American soil at the same time that it commented on modern America. The identification between poetry and technology, then, appeared to offer American poets a way of including the real world in their art and of defending their poetry in a culture that respected most the work of engineers or industrial designers, in short, the work of practical men. At the same time, the poets hoped to show European modernists that America's writers could be as inventive as its engineers.

The chameleon-like strategy of associating poetry with technology, engineering, or science posed problems, however. For one thing, to argue the importance of both scientific and poetic creativity to an American audience was unconvincing since most Americans equated scientific creativity with technology, including commercial products. Comparisons between science and poetry therefore suggested that poems were, like many technological inventions, commodities. Practically speaking (and speaking to an American audience requires speaking practically), it is difficult to explain the uses of poetry compared with the uses of engines or plumbing. For another thing, modernism in part was embraced by American writers because it broke with American genteel poetry, as George Santayana called poetry written in the tradition of Whittier and Longfellow. Yet although most Americans did not think genteel poetry relevant to daily life, dismissing it as irrelevant and even unmanly, they were not prepared to accept a new style. The general populace admired machines for their practical uses, not for their aesthetic qualities. The taste for an American vernacular style, or for modernist art and literature, was not widespread. Last, as the poets discussed here came to see, the very features of American modernity they wanted to appropriate for art defined a set of values in which the arts did not figure highly.

Chapters 1 and 2 fill out this picture. In particular, they examine the ways in which definitions of America and of an American aesthetic entered the public arena in the magazines, newspapers articles, and books American writers and intellectuals were reading and writing between 1910 and 1945. A number of writers—including the three poets on whom I will concentrate—show an acute awareness of the problems involved in defending a modern style in an American context. The early chapters of this study are concerned with illustrating how the ideas and problems with which Williams, Moore, and Stevens, struggled were part of current debates about business, commerce, the identity of America, modernism, and, above all, science and technology in America.

The science most often mentioned in early twentieth-century discussions of the modern age was technological. The century, however, also marked the beginnings of a scientific revolution in physics.[9] By 1921, when Einstein visited the United States, the physicist had become a folk hero and the new physics was front page news, as Carol Donley and Alan Friedman have shown.[10] The models of science presented by Werner Heisenberg, Max Planck, Albert Einstein, and popularizers like Alfred North Whitehead or Bertrand Russell, are different from the models of nineteenth-century science. Chapter 3 discusses how the new physics as popularly understood—and misunderstood—was related in the public imagination to American culture and to art and literature.

Einstein's original formulation of the special theory of relativity from 1905 states that whereas an event viewed by two separate moving observers may appear different to each, neither observer will be wrong, or encounter contradictions, if he or she uses the same basic laws of physics. For example, the speed of light is a constant. This might seem to lead to contradictions, since one person observing a light beam might be moving faster than another person observing the same light beam. What happens, according to Einstein, is that the nature of time and space are altered by motion; the laws of physics remain constant. Einstein's later work on general relativity then extended his ideas to cover curved time and gravitation.[11]

Max Planck's work also concerned light and motion, but concentrated on subatomic phenomena. In 1900, Planck discovered that electrons absorbed or emitted light in quantum units; he also found that there was a constant (h) by which to measure the values of such energy exchanges. His discoveries required abandoning the notion of a continuum of energy; Einstein later showed that Planck's findings suggested light was composed of particles. Einstein's paper on light particles or photons appeared in the same year that he published his special theory of relativity, in which light was treated as a wave.[12] Werner Heisenberg's 1927 work on the *uncertainty principle*, building on the work of Planck and Einstein, proposed that the error in position measurement times the error in momentum measurement can never be less than one half of Planck's constant. Position and speed cannot both be known.[13]

The new physics, then, broke down the framework of classical physics, suggesting that time and space were fluid and that phenomena changed depending on how they were observed (light being sometimes a wave, sometimes a particle, for example). Einstein, whose work on relativity reaffirmed the constancy of physical laws, resisted the implications of quantum theory. Nonetheless, most physicists agreed that the difficulty of defining light or measuring subatomic *wavicles* was not due to the failings of scientific instruments but to the actual nature of the physical universe, a universe of "fuzzy" statistical probabilities or, to use the term provided by physicist Niels Bohr, of complementarity.[14]

If the new physics changed ideas about the nature of the physical universe, popular accounts often misrepresented the implications of the scientific discoveries. Physics was related to democracy, to free will, to Bergsonian philosophy, and to literary experiments that toyed with perspective or emphasized motion.[15] Donley and Friedman argue that there are few "clear cut links that would establish a satisfying cause and effect between [the new] physics and [modern] literature."[16] Certainly, few laymen in the first half of this century read the physicists themselves and even fewer understood the new physics. That Einstein's resistance to quantum physics was in part an aesthetic response (as some historians of science now argue)[17] would certainly not have occurred to artists or writers in the 1920s, who often could not distinguish between Einstein's and Heisenberg's discoveries. Yet the new physics was popularly connected with an aesthetic of process and motion. Although James Joyce probably did not have Einstein in mind while writing *Ulysses*, Joyce's style was nevertheless equated with Einstein's universe in Wyndham Lewis's 1928 *Time and Western Man*, Edmund Wilson's 1931 *Axel's Castle*, Alfred North Whitehead's 1925 *Science and the Modern World*, and in reviews of these books and of the new physics in journals such as the *Dial* or *Hound & Horn*.[18]

Whitehead's book proved particularly important to the poets on whom this study concentrates. He explained that Romantic poetry had anticipated the new physics with its emphasis on the relationship between observer and observed and with its image of reality as an organic unity. Touching on the fields of philosophy and art, Whitehead proposed that the new physics demanded that mechanistic models of both mind and matter be replaced with models which stressed process and a creative engagement between mind and matter. Most important, Whitehead argued that one could be a objectivist without being a materialist, since the world was not composed of isolated objects or entities.[19] Whitehead thus suggested how the new physics offered a way to redescribe objects and motion—and so poetry—without invoking technology and without abandoning the claim that poetry was rooted in the physical world.

In general, comparisons between poetry and the new physics, like analogies between technology and poetry, served to defend the importance of poetry to an American audience that was as impressed by Einstein as by technology. As with references to technology, however, there were problems with the claim that Einstein had proven the validity of poetry; Einsteinian physics was, after all, held in higher esteem than poetry. To quote Donley and Friedman: "since Einstein's field of science was mathematical physics, that realm of endeavor is *the* prestige field in which to look for great intellectual achievement. The arts, poetry, politics, and even other sciences are perceived with less status, in part due to Einstein's eminence. We look to physical science for solutions to societies' problems with more seriousness, and more money, than we might if our culture had happened to choose a philosopher, a poet, or a psychologist for its symbol of intellect."[20]

References to Einsteinian physics by artists and writers served, in part, to propose that a poet or artist might equally well serve as the country's symbol of intellect. Yet the choice of Einstein as a symbol of intellect was not something that just happened; it was rather a choice fed by traditional American values. This is not to deny Einstein's true stature, but to suggest that his human image combined with the field of his achievements made him a modern myth. Americans' respect for science, in short, predated Einstein and was not wholly an accidental by-product of their respect for the man. Pasteur, a chemist, had a similar mythic status in the 1910s and 1920s in America. A national hero first in France, he, like Einstein, had the look of a friendly sage, which combined with the practical results of his work on aseptic surgery, rabies, and food decomposition (and his particularly dramatic role in saving children's lives) to make him an emblem of scientific creativity in the United States. It is beyond the scope of this study to trace the making of the Pasteur myth in detail. It seems, however, that, as with Einstein, Pasteur's fields combined with his genuine achievements to put him in the pages of *McClures* and the *Nation*.[21] Further, whether referring to Pasteur, American technology, or Einstein, many people—and in particular the three poets discussed in chapters 4, 5, and 6—recognized the difficulty of comparing poetry or poets with more valued fields and figures.

Nonetheless, the poets treated here do sometimes use analogies between science, technology, and poetry to grant authority to poetry by associating it with work in other fields perceived as more objective and more prestigious. Moreover, they were very much a part of American culture, sharing the American respect for what was useful or practical, for what was grounded in the real world. As Americans, then, the poets required a description of poetry's connection with practical reality—including the *real* world of science and technology—in order to defend the place of their art in modern society. At times, I suggest, the poets' questions about the relationship between poetry and these other fields obscured deeper questions about social values and institutions that devalued the arts. I will also suggest, however, that the poets themselves finally recognized this problem and that their pointed critiques of American society often grew out of their uneasiness with the use of references to or analogies with science and technology to defend poetry.

Williams, Moore, and Stevens are central to this study for two reasons. In the first place, of all the modernist poets who stayed in America between 1910 and 1945, they are among the most important as poets and also among the most interesting and influential as thinkers about the problematic role of poetry in American society. Hart Crane, who might otherwise be included here, left less of a legacy; his writings about poetry and its place in America (as opposed to his poems on that subject) are less widely cited by contemporary American writers.[22] In the second place, Williams, Moore, and Stevens provide exemplary case studies in what they reveal about American defenses of modernism and about how and why such defenses drew on the images and rhetoric of science and technology. These poets are original thinkers, but they developed their ideas while using and thinking about the commonplaces of their period in a critical fashion.

I have repeated some of the arguments made in chapters 1, 2, and 3 in the later chapters, in part to show the specific ways in which each poet encountered the ideas and images of the period and in part so that readers interested in a particular poet might be able to turn to individual chapters. Nevertheless, the book is meant to be read as a whole: just as these poets' careers give body to the more general context presented in the early chapters, so the American context illuminates the ideas and achievements of the individual poets. Similarly, an examination of one poet's work sheds light on the work of the others. Indeed, the order in which the poets are discussed is heuristic.

Chapter 4 begins with Williams because his interest in American culture and in science and technology is the most obvious. His exchanges with modern artists and educators, his definition of poems as machines, and his references to Einstein are widely commented on by critics and familiar to even the most casual reader of his work. In part because of his large number of interests, Williams also asks the most broad-ranging questions and helps clarify why American education, commerce, technology, engineering, architecture, and science are issues American poets addressed in the first four decades of this century. This chapter argues that Williams's attraction to a machine aesthetic, his connection with European modernists, and his career as a doctor helped him see how modernist defenses of a machine aesthetic were problematic in a country that appeared to value actual machines over aesthetics. At the same time, Williams's experimentalism and his fascination with a poetics of motion helped him realize how he might use the new physics to develop a new poetics that grounded poetry in the physical world without inviting American readers to compare poems with commodities. Williams's application (or misapplication) of physics reveals his respect for science. He equally valued the products of American technology, especially those that improved health care. Yet he did develop a critique of science, technology, and the way in which America overvalued both fields, a critique he did not abandon even in his use of Einstein to underwrite his mature poetics. Because Williams examines poetry's place in American culture so explicitly and so broadly, he helps clarify how and why Moore and Stevens also use images and language drawn from debates about America as a country whose strengths were perceived as being commercial and scientific.

Chapter 5 discusses Marianne Moore. Moore is less concerned with physics than either Williams or Stevens. She did, however, receive an undergraduate degree in biology and continued her interest in the natural sciences throughout her life. She also had a brief career teaching in a commercial college and a long-term interest in technology and industrial design. Like Williams, she admires not only the creativity of scientists but also the achievements of American technology and even of American business. Also like Williams, Moore sometimes draws parallels between poetry and other more accepted fields to grant authority to poetry.

Moore's usual strategy is to take the images and language used to describe American industry, technology, business, and science in debates of the period (many of which she published in the *Dial* while she was editor) and to redefine these commonplaces, pointing towards the creativity involved in all fields and, more generally, redefining the material world in spiritual terms. Even so, Moore shares with Williams and Stevens a characteristic American respect for facts, for the concrete, as well as for creativity. Her celebrations of both factual accuracy and creativity are not always easily reconciled, but more often maintained in a creative tension. Moore's poems and essays thus typically set a dialogue in motion, critically examine American language and rhetoric, and so embody the poetic thinking and attention to language she, like Williams and Stevens, believed central to poetry.

Wallace Stevens is discussed last, in part because his concern about poetry's social role is most clear when set in the context of Williams's and Moore's work. Reading Stevens in isolation, it is at first difficult to see that his defense of poetry is a response to the American context in which he wrote. Unlike Williams or Moore, Stevens rarely mentions the technological or industrial features of

America. Williams and Moore used images of modern America not only to reflect but also to reflect upon the nature and values of modern society; they would have agreed with Hart Crane's judgment that "To fool one's self that definitions are being reached by merely referring frequently to skyscrapers, radio antennae, steam whistles, or other surface phenomena of our time is merely to paint a photograph."[23]

Nonetheless, what the poets refer to reveals a good deal about the social and historical context of their work and their engagement with that context. Stevens's apparent silence about America and American technology in his poetry makes it easy to overlook his interest in these topics. Yet an examination of Stevens's journals and early drafts of some of his poems suggests that his very silence about American technology was a conscious choice, an indication of his studied objection to the ways in which others publicly defended poetry by comparing it to other fields that Americans found had more obvious utility. It is in this context, and set against Williams's and Moore's statements, that Stevens's response to a 1934 *New Verse* questionnaire is revealing. In answer to the question, "Do you intend your poetry to be useful to yourself or others?", Williams said his poetry was intended to be useful to both himself and others, though primarily to others; Moore, modestly, said her poetry could only be expected to be useful to herself.[24] Stevens, however, objected to the question: "Not consciously. Perhaps I don't like the word useful."[25]

If Stevens rejected the word useful, his other writings and his response to critics, especially during the Depression, indicate that he, like Moore and Williams, still sought a description of poetry's connection with practical reality. And like Williams, Stevens's founded his mature poetics on the new physics, which he believed rooted his long-term celebration of process and mental motion in physical reality. Chapter 6 traces Stevens's early attempts to define poetry in opposition to nineteenth-century science, his resistance to using characteristic features of modern industrialism to define an aesthetic, and his growing awareness of the uses to which he could put the new physics in his defense of poetry. Finally, Stevens's "act of the mind" (*CP* 240) is specifically an act he believed would help people live their lives and implicitly a definition of the importance of poetry in modern America.

Williams, Moore, and Stevens reveal the difficulties involved in defining poetry's place in a technological society. If using the images and language of science, technology, and industry to defend poetry sometimes posed problems for them, they are in many ways their own best critics in their awareness of, and responses to, these problems. Indeed, if the strategy of defending poetry by borrowing authority from other more valued fields persists, the more important legacy of the American modernists is also alive, namely their demonstration of poetry's ability to grant value to—and to think about—the language we use in all exchanges and in all parts of our lives.

1

The American Context: No Place for the Artist

THROUGHOUT the nineteenth century, American writers worried that American poetry was merely a pale reflection of English literature. Solyman Brown's 1818 "Essay on American Poetry," for example, calls for America's "emancipation from literary thraldom."[1] Anxiety about America's lack of culture and the need to define a place for literature in America were much in evidence in the century following, in spite of the accomplishments of Emerson, Hawthorne, Melville, Whitman, and Dickinson. In 1888 Charles Eliot Norton was still asking whether America had anything "to show in which the spirit of a great nation is revealed through . . . the works of its creative imagination?" Norton concluded pessimistically that in the United States "public opinion exercises a tyrannical authority . . . [and is suspicious] of independence and originality."[2]

Norton's comment is as much about the lack of public appreciation for art as about its production. American attitudes toward art, however, and especially toward poetry, had an important effect on those who aspired to be writers or artists in America between 1890 and 1930. By the 1870s, America had begun publicly supporting the visual arts. The Metropolitan Museum of Art in New York and the Museum of Fine Arts in Boston opened in 1870; the Philadelphia

Museum of Art was founded in 1876 and the Art Institute of Chicago in 1879. Such support for the visual arts may have involved the American search for aesthetic models with cultural authority. Yet, as Alan Trachtenberg has argued, private support for museums and for the visual arts generally was also a form of investment or conspicuous consumption; moreover, the museums "monumentalized the presence of culture," separating culture from the everyday lives of Americans.[3] By the 1920s, as noted in *Middletown*, the Lynds' 1929 study of the typical American community of Muncie, Indiana, there were community study groups on the visual arts; yet the Lynds were "struck by the gap that exist[ed] between 'art' as discussed in these groups and as observed in most . . . homes. . . . [Although] art [was] regarded, at least among the business class, as a thing of unusual merit, art in the homes [was] highly standardized and used almost entirely as furniture."[4]

Literature, and especially poetry, was equally removed from the lives of most Americans. In the typical American community, interest in literature seems to have declined in the first two decades of the century. The Lynds note that in the 1890s Muncie had reading circles, which organized discussions and recitations of work by Dickens, Burns, Ruskin, and others (although little of the poetry read even in these groups was by American authors), but by 1929 no such literary groups remained and even the visual arts were primarily of concern to the young, mostly to young women.[5]

Between 1900 and 1930 cultured readers turned to the high sentiments of American genteel poetry, like that of Trumbull Stickney and Thomas Aldrich, while the average American read the sentimenal verses of poets like Edgar Guest and James Whitcomb Riley in the leading magazines of the day, such as the *Saturday Evening Post*.[6] Although such poetry was viewed as a refuge from the harsh realities of the world, and so as having an almost sacred character,[7] it was not generally taken seriously. Nor were those who wrote poetry. The attitude voiced in a letter that accompanied some poems submitted to *Scribner's* is typical: A woman writes that her "husband has always been a successful blacksmith. Now he [is] old and his mind [is] slowly weakening so he [has] taken to writing poems, several of which [she encloses]."[8]

The insecurity of American poetry at the end of the nineteenth century, a period that E. C. Stedman called "a twilight interval," may in part be linked to what has been called the "feminization" of poetry, though it may also be that the so-called genteel tradition of the late nineteenth century was itself a failed attempt to defend an already insecure vision of poetry.[9]

The image of writing poetry as an unmanly and less than serious pursuit is repeated often through the late 1920s even by poets. In 1913 Stevens asked his wife to keep secret his attempt to put together a collection of poems, confessing he found "something absurd about all this writing of verses" (*L* 180). Williams wrote to his brother after the publication of his first volume of poetry in 1909, "A good many people think to like poetry is to be a molly coddle."[10] Williams's early poetry was, as his wife admitted in a 1962 interview, "pretty bad,"[11] yet even after he had written some of his best poems, he was still painfully aware of American attitudes towards poets. In the late 1920s he noted that there was "a subtle loss of dignity in saying a man is a poet instead of a scientist" (*EK* 51), and as late as 1949 he was still saying: "We seldom think of . . . poetic structure, as we do of engineering: a field of action worthy of masculine attack."[12] Williams was not alone in feeling that poetry was thought of as unmasculine and was less valued than either science or engineering. Walter Pach, for example, writing in 1922, echoed Stevens's and Williams's regrets that in America art, as opposed to business or engineering, "carrie[d] with it some implication of effeminacy."[13]

Pronouncements about the effeminacy of literature involved an awareness both of the inadequacies of the genteel tradition and of poetry's association with the private, rather than the public, realm. Thus, as Williams's and Pach's references to engineering also suggest, attitudes towards poetry were related to larger social issues. Williams's and Pach's statements in fact responded to a widely shared analysis of American society and culture. *Civilization in the United States,* in which Pach's essay appears, is a diagnosis of American society by a number of writers, including Van Wyck Brooks, Lewis Mumford, and H. L. Mencken, who had been publicly pursuing the problem of how to reinvigorate American art and the quality of life in America for over two decades.

For Brooks, the lack of a shared, indigenous culture devalued both literature and daily life in America. In a 1908 article, "Harvard and American Life," Brooks criticized American education for resembling industrial America at large. In particular, he charged that his alma mater had become "the factory of . . . imperialism" and a center for the production of business men.[14] Brooks argued that "the American type more and more define[d] itself" in the new educational system, and "with the University the efficient practitioner of the future emerge[d]."[15] In *The Wine of the Puritans*, also published in 1908, Brooks widened his focus and condemned the spiritual emptiness of all American experience, past and present.[16]

The Wine of the Puritans presents an essentially pessimistic critique of America; Brooks leaves his readers with no concrete suggestions about how to remedy the ills described. There is a slightly less sweeping analysis of American history in the article on Harvard, in Brooks's 1915 book, *America's Coming-Of-Age*, and in his 1922 essay in *Civilization in the United States*. In these writings, Brooks emphasizes the industrialization of American society as a source of the impoverishment of American life, education, and art. Thus, in 1922, he concentrated on the decline of American literature since the 1870s, the first decade after rapid industrialization and urbanization began in America.[17] Brooks concluded that "as one surveys the history of our literature during the last half century, [one notices] the singular impotence of its creative spirit."[18] Brooks saw Puritanism and the lack of a holistic, medieval tradition as America's problem, but he also added, more convincingly, that modernization had caused a deterioration in the quality of American art and American lives. Brooks's view that progress in business and industry had resulted in a split between a spiritually empty work life and an equally bankrupt culture was shared by others, many of whom published their views in the *Seven Arts* and in the *Dial*.[19]

Many members of the *Seven Arts* circle (which included Brooks, John Dewey, Randolph Bourne, Harold Stearns, Lewis Mumford, Leo Stein, and Waldo Frank) and those affiliated with it (such as Alfred Stieglitz, George Santayana, and Hart Crane), believed the arts should be made relevant to the lives of common Americans, in part to give value to their spiritually empty existence. Although many poets had reservations about some of the *Seven Arts* group's ideas, they often agreed that common men and women should be their audience. In 1917, for example, Williams wrote: "All this— / was for you, old woman. / I wanted to write a poem / that you would understand. / For what good is it to me / if you can't understand it? / But you got to try hard" (*CEP* 165–66). The colloquial speech as much as the statement underlines the desire to write for the woman on the street, a woman like many of Williams's patients.

For Brooks, as well as for Williams and Moore, the conception of the common American as the audience for poetry was important because providing Americans with a way to reevaluate their everyday lives was seen as one of poetry's functions. Further, many writers shared the assumption that the practical inventiveness of common citizens was a sign of America's creative genius. By the end of the nineteenth century, however, this assumption was, in part, myth. Edison, for example, was seen as a typical American craftsman-inventor, and he promoted the image of himself as a man of instinctive creativity, without formal training. Yet, in reality, as Trachtenberg suggests, Edison's genius was for business. Instead of relying on the inventiveness of craftsmen and entrepreneurs like those who had spawned the growth of American industry, Edison turned to an emerging class of specialists required by industry's rapid growth and employed university-trained scientists.[20]

Edison's image making appealed to a widespread popular mistrust of the growing professionalism and bureaucratization of both industry and the academy.[21] In various hands, the celebration of the amateur was linked with American anti-intellectualism and with a belief in populist democratic principles.[22] A full consideration of these attitudes in American intellectual culture is beyond the scope of this book, so I will only mention a few of the ways in which they affected attempts to define the place of literary culture in America.

Van Wyck Brooks's position in 1915, in *America's Coming-Of-Age*, is a good case in point. By 1915, Brooks described America as "an unchecked, uncharted, unorganized vitality" that needed to be "worked into an organism."[23] Brooks then argued that American vitality might be used to renew the lives of common Americans and reintegrate culture and work. Yet even as he described the organic community he believed would cure American ills, Brooks could not imagine the common man in America creating the culture he desired, nor did his ideas take account of the actual beliefs of most Americans. In short, Brooks's criticisms of business and commerce were not likely to appeal to the man on the street, and he was left with little place to turn for the organic community, with a place for art and human values, that he envisioned.

Brooks's position was in some ways unique, but nonetheless it reveals the problematic nature of the stand taken by many other defenders of American literary culture, especially of American poetry. Williams and Moore, for example, defended what they took to be democratic positions and acknowledged their own suspicions of certain kinds of expertise. Williams's identification with the common man and woman, in poems like "Proletarian Portrait," for example, is also linked with his fierce anti-intellectual and anti-academic stand in books such as *The Embodiment of Knowledge*, from the late 1920s, where he criticized would-be experts in science and philosophy. Moore also wrote of America in "The Student": "We / incline to feel, here, / . . . it may be unnecessary / to know fifteen languages" (*Comp* 101), and she noted the dangers of academic life: "bookworms, mildews, / and complaisancies" (*Comp* 102). And although ordinary Americans found her early poetry obscure, when asked what distinguished her as a poet from the ordinary man, Moore responded, "Nothing."[24] The difficulty of such postures is underlined by Stevens, who answered the same survey question as Moore by saying he had an "inability to see much point to the life of an ordinary man" but could also say that he himself found the magazine Moore edited to be too effete; the *Dial*, Stevens judged, was "literary foppery."[25]

Brooks, Stevens, Williams, and Moore, then, shared a suspicion of poetry distanced from practical experience, yet their own writing was not accessible to most Americans. Williams's statement—"I need a friendly audience" (*SL* 224)—was not simply a confession of a private desire for approval. Rather, many writers believed poetry was validated by its ability to "help people to live their lives" (*NA* 29); to do so, poetry needed an audience. In his mistrust of the people whom he implicitly took as the community he served and the community whose attitudes he often assumed, Stevens's position is close to Brooks's.

Williams and Moore faced a slightly different problem: they claimed to be and to depend on ordinary Americans but were forced to recognize that ordinary Americans might not claim them. As Horace Gregory pointed out in 1946, "Dr. Williams seems to be speaking for [the common reader] in 'little magazines' that the 'common reader' never reads."[26] The common reader, if he or she read poetry at all, was reading Edgar Guest and the *Saturday Evening Post*. In short, efforts to claim that poets provided America's collective voice and that poetry was a public service were strained.

The so-called Little Renaissance of 1914 to 1918 was the occasion for many optimistic statements about how poetry would be reborn in America because the public needed poetry, even if they did not yet know they needed it. It is worth considering the story told on the mastheads of the magazines that devoted themselves to revitalizing poetry. *Poetry Magazine* borrowed its motto from Whitman: "To have great poets there must be great audiences too." The founder of *Poetry*, Harriet Monroe, hoped she could help create that audience. Yet she was also practicing a hopeful flattery; her propectus for the magazine acknowledged that she was catering to a small elite, whereas in general in America there was "a large public little interested in poetry."[27]

Similarly, in the 1920s, the *Dial* looked back to the age of Thoreau, Emerson, and Fuller, envisioning itself serving a wide public. Admirers of the *Dial's* accomplishments between 1920 and 1929 convincingly remark on the quality of work published there, including not only essays by Bourne, Dewey, and Bertrand Russell but also some of the best poetry being written in America. It is ironic, however, to find a recent essay idealizing this golden decade of the *Dial*, mourning that today "there is no obvious middle ground between the arts magazine and the general publication."[28] Robert McAlmon's letter to Stevens about the magazine, while it may represent in part a private grudge, still suggests that middle ground was not so easily found in the 1920s; McAlmon declared the "Dial group are all snobs."[29] Henry McBride's judgment of Stieglitz's position in America is also applicable to the magazine that published that judgment. McBride noted that even as Stieglitz's ideal was to produce an art for the masses ("numberless prints . . . circulate[d] . . . at a price not higher than that of a popular magazine, or even a daily newspaper"), Stieglitz was actually producing something "essentially aristocratic and expensive" in an "age of snap-shotting."[30] In short, poets, artists, and reformers such as Moore, Stieglitz, and Brooks wished to draw on the experiences of the common man, hoping to fill what they saw as the needs of the American populace, but the people of America as a whole did not develop a taste for the new in art or poetry.

The *Little Review* was more honest, in ways, than *Poetry* or the *Dial;* its motto read: "a Magazine of the Arts Making No Compromise with the Public Taste." The *Little Review,* however, ignored the fact that it contained work by Americans wishing to serve the public. Still, the motto acknowledges what even Williams had to grant, as in his pronouncement over the demise of the magazine, *Others:* "America has triumphed!" (*SL* 33). Indeed, in 1962, after Williams had published forty volumes and just before he was posthumously awarded a Pulitzer Prize, his wife remarked that "to this day very few people in Rutherford [the town in which Williams lived all his life] know anything about Bill's writing. . . . [He] has never attracted a general audience."[31] Like *Others,* many small magazines stumbled over the problem of whether American art could be both American and well received.[32] Adopting ideals and suspicions deeply engrained in popular American culture, those who self-consciously proclaimed themselves spokesmen of the American experience often found themselves appearing to have, and to speak for, no community at all.

Even those who followed modern developments in the literary world and recognized the inadequacies of genteel poetry did not pay much attention to the work being published by Williams, Moore, and Stevens. In 1938, when Stearns published a companion volume to the 1922 *Civilization in the United States,* John Chamberlain judged that the "'great writers' whose absence Van Wyck Brooks had deplored seemed more absent than ever," while Louise Bogan, acknowledging that a brief flurry of poetic activity had occurred in America from 1912 to 1918, pronounced that art was again comatose.[33] Part of the pressure on American poets in the 1920s and 1930s, then, was not only to relate poetry to public reality but also to convince Americans that they needed a new kind of poetry. This put poets in a logical bind. Poetry and the arts were called upon to reintegrate Americans with their lives and, by doing so, to forge a national culture. Yet the Americans whom the poets hoped to reach had little use for innovative poetry or art.

There were further dilemmas generated by analyses that saw the split between technological society and high culture as a major part of social and artistic problems. In order to outline the difficulties with such analyses, it is first necessary to understand how modernity, technology, and science were defined and how they were sometimes used as interchangeable terms.

Modernity was identified in a number of ways. Brooks's 1908 criticism of Harvard's efficient, factory-like production of businessmen points toward a number of changes in modern society: the rise of Frederick Winslow Taylor's techniques of "scientific management" (widely known by 1912); the efficiency not only of machines but of assembly lines (much in evidence by 1915); and the practices of American business and trade, which treated people as if they were manufactured, commercial products.[34] The same metaphor of production, linking efficiency, mechanization, commerce, and dehumanization, appears in John Dewey's 1917 assessment of "shop efficiency" in schools and his 1918 judgment that American education was being eroded by the "disintegrating tendencies of contemporary industry and trade."[35] In 1934, in *Exile's Return*, Malcolm Cowley described America's machine culture as the reason why many American artists and intellectuals left for Europe in the 1920s; art, Cowley concluded, could not flourish in a place of "efficiency, standardization, mass production, [and] the machine."[36]

The same features of industrial America were sometimes described as scientific. Taylor himself gave the name "scientific management" to his techniques for labor management, and Dewey similarly makes a connection between mechanical efficiency and the "modern scientific spirit."[37] Both Dewey and Taylor fail to distinguish technical from substantive reason[38] and they identify the institutions of a technological society with scientific thinking. A similar failure to distinguish scientific reason, technological management, and commercial products is found in many writers between 1900 and 1930. Thus, in *Hound & Horn* in 1928, John Walker reviewed a biography of Henry Clay Frick and wrote of Frick's part in the steel industry: "the 'scientific age' has . . . completely changed world conditions."[39]

The multiple ways in which *science* was understood can be detailed by looking closely at I. A. Richards's 1926 *Science and Poetry*. Richards taught English poetry at Cambridge University until 1939, when he moved to Harvard. *Science and Poetry* was one of a series of books in which he attempted to articulate the principles of literary criticism and to demystify poetry by providing a practical method that common readers might use in reading. Richards explicitly proposed a scientific method of reading and drew on positivist thought in his anti-metaphysical bias.[40]

In *Science and Poetry*, Richards asks, "What reasons are there for thinking [poetry] valuable?" (p. 9). Although committed to scientific methods in his own approach, Richards argues that "poetry is just the reverse of science" (p. 24) and that truth or knowledge of the sort science offers is irrelevant to man's aims (p. 52). The poet, Richards continues, gives "order and coherence, and so freedom, to a body of experience" (p. 55). The poet's business is not scientific truth ("ultimately a matter of verification as this is understood in the laboratory"); rather, poetic statements are "pseudo-statements," validated by their ordering effect on a reader (pp. 56–60). Richards thus struggles to separate science from poetry and the criteria by which scientific results are judged from those applied to poems.

Although Richards wants to separate poetry and science, he defines the effects of poetry (which for him are the basis of its value) mechanistically. He likens a reader's response, for example, to a "'governor' run by but controlling [a] main machine" (p. 15). In several ways, then, Richards uses science to grant authority to poetry, even as he explicitly sets out to defend the importance of poetry in a scientific age. First, for Richards, science is the final arbiter of truth, in that he justifies his view of what poetry is by appeal to the "*science* of psychology" (p. 41, emphasis added; also see p. 55 and passim). Second, he implicitly defines science not only as a rational method of inquiry but also as mechanistic.

As R. P. Blackmur suggests, Richards turns aesthetics (etymologically, the art of seeing) into the mechanics of perception. Blackmur's first objections were published in 1928: "If, as many think, science has erased many of our intellectual convictions, and if, as I. A. Richards asserts, beliefs are hardly possible, it is only because either our convictions were skin-deep and not worth holding, or else we have allowed science to upset matters with which it can have no legitimate contact."[41] Blackmur wants "to restore the intelligence and the sensibility by adverse criticism of any such misapplications of science."[42] In short, for Blackmur, humanistic and artistic endeavors have to be distinguished from the mechanical and theoretical sciences in order to defend the world of values and perhaps in order to defend the arts as valuable in a modern world.

Blackmur's statements help to uncover some of the characteristic motives behind the various kinds of allusions to "science" in Richards's work. Blackmur charges that Richards's scientific method is not a "legitimate" way to judge poetic truth and, further, that perception or aesthetics cannot be mechanized. Although Richards's use of a scientific method is intentional, his references to mechanics and to other applied sciences appear more confused. For example, in his introductory chapter, Richards discusses the practical results to which scientific discoveries can lead, reminding his readers that only "persons thought to be crazy knew before 1800 that ordinary traditional ideas as to cleanliness are dangerously inadequate. The infant's average 'expectation of life' has increased by about 30 years since Lister upset them" (p. 2).

Lister, the English surgeon, collaborated with Pasteur, the French chemist. As early as 1902 a biography of Pasteur was available in English, and some popular magazines also emphasized Pasteur as a research scientist. In one example, an anonymous reviewer for the *Nation* commented on Pasteur's "childlike insight into things" and his attention to details without an eye to results.[43] As a doctor, Williams knew of Pasteur's work; in fact, in a 1928 essay he mentions the French chemist as a young scientist whose work was experimental (*I* 365).

In effect, Richards suppresses the theoretical nature of Pasteur's work in bacteriology and concentrates instead on the practical applications of theory by Lister (in part a bow to the English portion of his audience). Richards does describe Lister's accomplishments in terms of his creative overturning of traditional ideas, but he also implies that traditional ideas are upset when confronted with visible, empirical evidence. As Lionel Trilling has argued, "in an age of science prestige is to be gained by approximating the methods of science."[44] Trilling is referring to scientific method, but Richards's own discussions of science reveal that he is explicitly drawing on the prestige of the applied sciences.

Leo Stein—Gertrude Stein's brother, a founding member of *The Seven Arts,* and an important collector and patron of modern art—similarly elevates practical or applied science in a 1917 article, "American Optimism." Stein begins by separating science and technology, yet ends by arguing that progress in the arts has lagged behind "scientific" progress. Stein then describes scientific progress in terms of practical results when he calls for men in the arts to produce changes comparable to those produced by Lister and Pasteur.[45] Like Richards, Stein implicitly measures progress in terms of technological progress or visible results. Although Lister and Pasteur demonstrate that there can be a close connection between applied and theoretical science, both Stein and Richards confuse the two as a way of making their arguments work.

A particularly telling example of this confusion of science with technology is found in Richards's conclusion to his introductory chapter where, perhaps unconsciously, he mixes discoveries in theoretical science with the results of applied science in order to defend his own method of reading (which was based on the science of psychology) and also to defend poetry, which he claims might save humanity (p. 82). He argues that "if only something could be done in psychology remotely comparable to what has been achieved in physics, practical consequences might be expected even more remarkable than any that the engineer can contrive" (p. 6).

The shift from physics to engineering allows Richards to emphasize the practicality of his own enterprise; the effectiveness of the strategy rests on Richards's and his audience's definition of engineering as both practical and comparable to theoretical science. The argument might be rephrased as follows: if psychology could be revolutionary like other sciences (physics), then it could produce results like other sciences (engineering). Throughout *Science and Poetry,* science is variously defined as scientific method, mechanics, industrial machinery, hygiene, engineering, and physics. By conflating pure science with technology or applied science, Richards can compare the creative features of pure science with those of poetry and then claim for poetry the strengths and prestige of applied science.

Such a sleight of hand was important, especially in America, where industry and business were equated with technology, since applied science had prestige and theoretical science did not. Indeed, Robert H. Lowie's chapter on science in *Civilization in the United States* argues that science must not be subject to the "principles of business efficiency" and suggests that science itself felt as constrained by the American climate as poetry did.[46] As Alan Trachtenberg has pointed out, "in America, . . . formal science bore the onus of impracticality and remoteness from human need."[47] Jackson Lears also cites the publisher Henry Holt's 1895 pronouncement about the blindness to social circumstances found in "the large majority of intellectual people . . . peacefully sitting reading physical science and the classics."[48] Moreover, those such as Charles Eliot Norton who did not wish to see the loss of "old-fashioned literary culture" also mistrusted science, which seemed to them increasingly too professional.[49] One problem, then, with comparisons between science and the arts was that the traditional status of pure science was not what the arts would want to appropriate. And Richards's and Stein's slippery definitions of *science* to defend other disciplines were also ironic, since science as popularly understood, which is to say technology, appeared to be why the study and writing of poetry and the production of art were devalued.

The negative association of science with mechanization, labor management, technology, and practical results is particularly found in the early twentieth century. The idea, however, that science and the philosophical changes that paralleled the rise of science threatened spiritual and artistic values has its roots in the late eighteenth century. The English Romantics provided the most influential critique of poetry's relationship to scientific and empirical knowledge, a critique that, in modified form, persists to this day. M. H. Abrams argues, in *Natural Supernaturalism*, that the "Romantic enterprise was an attempt to sustain the inherited cultural order . . . to save what one could save of [Christianity's] experiential relevance and values."[50] That is, Romantic poetry became the repository of values that empiricism and science ignored as irrelevant to everyday experience.

Poetry, as the Romantics from Blake to Coleridge insisted, became the refuge for higher human values and for a holistic, imaginative vision of the world. For Shelley, poetry "is that which comprehends all science, and that to which all science must be referred."[51] Although it is Mary Shelley's *Frankenstein* that presents the clearest example of the century's fears about science and its productions,[52] Wordsworth, Coleridge, and Shelley all reveal a need to defend poetry against science, and to explain poetry's worth to a skeptical world. Shelley's *Defense*, for example, was provoked by Peacock's conclusion in *The Four Ages of Poetry*. Peacock wrote that poetry's cultivation "must necessarily be to the neglect of some branch of useful study," with science and philosophy as the examples of such study.[53]

Gerald Graff has noted that "romantic aesthetics typifies the more general crisis of modern thought, which pursues a desperate quest for meaning in experience while skeptically unable to accept the validity of any meaning proposed."[54] Michael Fischer offers a parallel analysis of Romanticism when he suggests that "Romantic writers . . . used the imagination to authorize those value judgments which they and their mechanistic adversaries placed beyond the grasp of rational knowledge. This degree of complicity with their opponents, however, undermined their own confidence in the truth of poetry."[55] That is, in an age where science provides empirical truths—such truths having practical and monetary value in technological applications—science and the arts may agree that value judgments are not the province of science. Yet problems arise when science and the arts also agree that what is not scientifically true or practically effective is not valuable. Understanding this, Stein discusses how art must produce results, Richards places reading on a scientific basis, and Brooks and Dewey try to reintegrate art and practical experience.

Cultural critics such as Brooks and Dewey often identified the separation of art from practical experience as symptomatic of the larger problem the Romantics first encountered, namely the exclusion of values from the practical world of experience, because their primary concern was with social problems.[56] On the other hand, for those whose concern was art or poetry, the question still remained: How might poetry be tied to practical American reality and how might the function of poetry in America be defined practically? Although they held different positions, for both Williams and Stevens the issue of how to connect poetry with quotidian reality was important because of their desire to establish a place for poetry and poets in America. In 1922, addressing just this issue, Wallace Stevens asked whether "man is the intelligence of his soil" (*CP* 27) or "his soil is man's intelligence" (*CP* 36). Crispin, the comedic anti-hero of "The Comedian as the Letter C," decides the latter to the detriment of his poetry.[57] And Williams explained his search for a new poetry by citing his practical experience as a doctor and as an American: "It may have been my studies in medicine; it may have been my intense feeling of Americanism; anyhow I knew that I wanted reality in my poetry."[58]

Yet, as with Richards and Stein, those who desired to include practical reality in poetry faced the problem that modern America seemed to have no place for poetry. Following Brooks and identifying factory work and a mechanized environment with consumerism and materialism, Hart Crane's 1920 "Porphyro In Akron" examines a "shift of rubber workers" in a city whose "axles need not the oil of song." The poem concludes that in industrial America "poetry's a / Bedroom occupation." Commercial as well as technological modernity is further identified by Crane as the reason America held the arts in low esteem and relegated poetry to the private sphere. Crane's Americans are only interested in using "the latest ice-box and buying Fords."[59]

Technological and commercial modernity was particularly visible in America, where products such as vacuum cleaners, enameled bathtubs, telephones, and automobiles had a deep effect on people's daily lives;[60] such products also were commercially available in America sooner than elsewhere. Perhaps for this reason, the association of materialism, commerce, technological success, and efficiency with American modernity was widespread. In his 1922 review of *Civilization in the United States,* George Santayana perceptively comments that Brooks and other contributors to the volume had not noticed that "the whole world is being Americanized by the telephone, the trolley car. . . . Americanism . . . is simply modernism—purer in America than elsewhere."[61] Yet even Santayana calls the process of modernization "Americanization."

In fact, America was not the only modern country between 1900 and 1920; in their earliest writings, both Dewey and Bourne pointed out that Germany, for example, was at least as efficient as America.[62] Nonetheless, modernity was identified as American, and it was judged responsible for America's lack of culture.

Despite their misgivings about the effects of modernization, many poets and writers included what they saw as the distinctive features of America in their work, in part to make their writing relevant to their American audience. They saw that literature, to be taken seriously, had to be related to what were seen as the values of modern society. At the same time, however, they feared that Americans could not be weaned from their admiration for commerce, business, and mechanization. Thus Crane, who chose a major feat of engineering as the center of his American epic and wrote that poetry had to "absorb the machine" in order to serve "its full contemporary function," had moments when he felt the technology he sought to revalue could not be used in poems: "The bridge as a symbol today has no significance beyond an economical approach to shorter hours, quicker lunches, behaviorism, and toothpicks."[63] And in his 1915 *America's Coming-Of-Age*, Brooks worried that not one poet "had the power to move the soul of America from the accumulation of dollars."[64]

The question, then, might have been whether poetry could be empowered by referring to the material features of urban America when the materialism of America's technological society threatened to eclipse all other values; yet this question was not often asked. For example, James Oppenheim, citing Dewey and Bourne, wrote that America had "no folk, no soil song or literature" in an article entitled "Poetry—Our First National Art."[65] Oppenheim's proposal for how to "produce a truer and more American art" was that poets should record "the direct impact of environment" (p. 242); however, he gives only vague explanations of how he thinks poets should register this impact or produce an art that is "true." Oppenheim implies he wants poetry to be not only more truly American, but truer also in the sense of encompassing reality. Some examples of reality that he mentions are "American speech" (pp. 240–42) and "environmental colour, snap-shots," which he finds in the work of Carl Sandburg and Walt Whitman (p. 239). Except for the reference to Whitman and American speech, Oppenheim's article suggests why Stevens might have objected to the emphasis on localism, which appeared to involve a static conception of reality.[66]

Oppenheim's reference to Whitman probably draws on Brooks's earlier description of Whitman as the one poet who, while he did not change American attitudes, "effectively combine[d] theory and action . . . [having been] a great vegetable of a man, all of a piece in roots, flavor, substantiality, and succulence, well-ripened in the common sunshine."[67] Although readers of Whitman give us many different versions of the poet,[68] Brooks and Oppenheim champion him for the way his poems are rooted in American experience; in particular, Whitman's poems give value to the technological reality of America. For example, the 1871 "Song of the Exposition" describes the muse:

> By thud of machinery and shrill steam-whistle undismay'd,
> Bluff'd not a bit by drain-pipe, gasometers, artificial fertilizers,
> . . . .
> She's here, install'd amid the kitchen ware![69]

Whitman's muse lives amid the products—such as machines, technological products, and domestic appliances—that testify to American inventiveness. As Leo Marx argues, Whitman's admiration for machines is genuine, and of all nineteenth-century American poets Whitman "comes closest to transmuting the rhetoric of the technological sublime into poetry."[70] Moreover, Whitman often focuses not on machines themselves, but on the intelligence and creativity of American inventors. In "Song of the Broad-Axe," for instance, he writes that the products of manufacturing and engineering "are not to be cherish'd for themselves."[71]

"Passage to India" (also from 1871) is Whitman's purest celebration of technological progress and yet, as Leo Marx again argues, even in "Passage" Whitman betrays a perhaps unconscious fear that industrial America might not be easily reconciled with Whitman's Edenic, spiritual vision of the United States; thus, the rhetoric in the poem lacks the redeeming particularity of Whitman's greatest poems.[72] Whitman's conscious attempt in "Passage" is to praise the spiritual energy evident in the achievements of nineteenth-century American engineering. The poem also explicitly redefines the American spirit as poetic; the apotheosis of the work of American engineers and scientists will be in American poetry:

> After the great captains and engineers have accomplish'd their work,
> After the noble inventors, after the scientists, the chemist . . .
> Finally shall come the poet worthy that name,
> The true son of God shall come singing his songs.[73]

On the one hand, Whitman sees poetry as the voice of technological and scientific America, and so provides a model of how to marry culture and practical reality for Brooks and Oppenheim. On the other hand, in the passage just quoted, Whitman must subordinate the accomplishments of American technology to define a place for poetry in America.[74] "Passage to India," then, is defensive about American technology even as it appropriates recognized American strengths for poetry and proposes that engineers and poets share a common muse.

To read Whitman in this way is, of course, to slight other features of his poetry. Yet it is important to see that the problems faced by writers who wished to connect poetry with practical reality in the early twentieth century are already foreshadowed in the poetry to which they looked for a model. Most significantly, Whitman shows the difficulty of including public accomplishments in poetry when America takes those who have physically built the distinguishing features of the modern world—businessmen and engineers, for example—more seriously than it takes poets.

Whitman's celebrations of engineering in the 1870s may already have betrayed a suspicion of how easily poetry was overshadowed by industry and science. Certainly the values of America a half century later confirmed that suspicion; even admirers of Whitman like Brooks saw that Whitman had not changed "the soul of America."[75] If anything, the soul of America, according to Brooks, Crane, Santayana, and others, was doing less well than it had been in Whitman's day. By the 1920s, the everyday reality of America was not only industrialized but also, in the view of many, utilitarian, materialistic, and commercial. It was, then, increasingly difficult to defend poetry's place in an American context. Although poetry was described as a cure for the ills of modern alienation, Americans appeared more inclined to look for material solutions to problems and to focus their attention—as well as their money—elsewhere. In sum, in consumer America, poetry had difficulty competing.

Poets and artists often specifically blamed commercialism, fed by mass production and advertising techniques, for their failure to reach a public. In 1921, the same year he left America, Marsden Hartley published "Modern Art in America," in which he wrote: "Art in America is like a patent medicine, or a vacuum cleaner. It can hope for no success until ninety million people know what it is."[76] Hartley was not alone in his disapproval of treating art like a commodity, a product of technology, that must be sold. Nonetheless, those who did not leave the country had to address the American predilection for buying vacuum cleaners rather than art.

In the late 1920s, Williams, for example, said he was amazed "that with all the cash there is free in America there is not one great mind with genius . . . to endow, or buy work of the rare few who are doing modern work" (*EK* 120). Similarly, during Moore's tenure as editor for the *Dial*, the magazine described its 1927 award to Williams in language underlining the connection between American commercialism and the lack of support for American poetry: "*The Dial* is in a way . . . diminishing, by a little, the discrepancy between [a writer's] minimum requirements as a citizen in a commercial society and his earnings as an artist."[77] And in his humorous 1921 article in the *Dial*, in defense of Moore's and Williams's poetry, W. C. Blum commented that "Williams is one of the people who think they know what the U.S. needs in order to ripen a literature."[78] The image of ripening was taken from Brooks's description of Whitman, and Blum then parodied Brooks's and others' suggestions about how to revitalize poetry: Brooks and the rest of the *Seven Arts* circle were too spiritual; Pound wanted everyone to read the classics; and Williams believed someone should give Alfred Kreymborg (an editor and writer) one hundred thousand dollars. Blum's irony is evident, but so is his approval of Williams's position. Moreover, Williams's own statements suggest he realized America would award poetry one of its highest signs of approval if it were to give money to an arts advocate like Kreymborg.

Few writers hoped to support themselves as artists. The modernists did believe, nevertheless, that poetry had something to offer people. To convince an American audience of poetry's value, their defenses of their art often invoked common American values—practical success, utility, efficiency, and even disdain for poetry—in order to reach their audience. At the same time, the poets attempted to undermine the more familiar American attitudes they included in their poems.

Moore's longer version of "Poetry" (*Comp* 266–67) is a masterpiece of this rhetorical strategy. She assumes the voice of America confronted with "Poetry," and agrees, "I, too, dislike it." Moore adds that poetry does, however, have "a place for the genuine." She then lists a series of images of physical reactions, which are to be valued "not because a / high-sounding interpretation can be put upon them but because they are / useful." Nor, she writes, should poetry exclude "business documents and / school-books." Moore defines the genuine in terms of human responses to the world and further notes that anything can be raw material for the imagination as long as it is looked at "in / all its rawness" and lifted out of its conventional setting. Poems, she concludes, are "imaginary gardens with real toads in them." At the same time that she redefines poetry, Moore also redefines the language of American advertising: "genuine" and "useful" are words most commonly used in advertisements for appliances in the period. Again, Moore adopts the language and attitudes her audience knows and believes in, while subtly resisting the usual definition of authenticity as relating to a trademark or to the usefulness of a labor-saving device. The real toads in Moore's poem are not the images she uses so much as the language of everyday America.

Williams, too, incorporates and challenges American attitudes. The 1923 *Spring and All* opens with a challenge to those who say, "I do not like your poem" (*I* 88). The dismissal of poet and poetry is repeated in poem XXV: "Somebody dies every four minutes / in New York State— // To hell with you and your poetry—" (*I* 146). The prosaic fact, both a fact of life and a statistic, that opens the poem reflects the language of contemporary American life brought to self-consciousness in the poem. In one sense, the American mistrust of poetry—the feeling that poems are not matters of life and death—is not overtly challenged in poem XXV. But by relating the contempt for poetry to a statistical approach to life as well as by lifting the speech of America into his poem, Williams tries to subvert American attitudes.

"Poetry" and *Spring and All* were not accorded the audience they would have required to change the lives of Americans. Both poets continued to worry about the lack of an audience for their work; in fact, both "Poetry" and *Spring and All* were first published outside of the United States, in part because American publishers did not support experimental work. Further, like Stevens's description of poetry in the same period (between 1910 and 1930), Williams's descriptions of his practice were not always as accomplished as his poems. Indeed, as they developed their poetics, and their poetic style, all three poets paid attention not only to debates about American modernity but also to debates about literary and artistic modernism.

2

Modernism, Modernity, and Technology: Following the Engineers

WILLIAMS, Moore, and Stevens all followed the debates about how to revitalize American poetry and modern culture. Williams, for example, responded to John Dewey, especially in *The Embodiment of Knowledge*, a book on education, science, and philosophy written in the late 1920s, but not published in Williams's lifetime. As I argue in more detail in chapter 4, Williams both disagreed with and learned from Dewey's program for integrating inventiveness and practicality. Similarly, Williams used and redefined Dewey's statement that "locality is the only universal" (see *SL* 224).[1] Moore's notebooks and *Dial* editorials reveal that she too read and responded to books and articles by Harold Stearns, Leo Stein, George Santayana, Randolph Bourne, and Dewey. Stevens was quieter about his interest in how American poetry and society were publicly defined, but in his 1922 poem, "The Comedian as the Letter C," he specifically takes up the arguments and vocabulary of Dewey and Oppenheim.

Some of the writers who were publishing in journals like *Seven Arts* were at the same time affiliated with Alfred Stieglitz's art gallery, 291, and thus with a group of painters and photographers who were doing experimental work in the visual arts. The artists who exhibited in Stieglitz's gallery formed a separate group from the writers and they, in particular, helped to make the poets aware of experiments in the visual arts and of the international movement in art and architecture known as modernism.[2]

Williams, Moore, and Stevens were certainly aware of these developments. They published in small magazines, such as *transition*, associated at home and abroad with the modernist movement. They knew the work of modernist writers in Europe such as James Joyce and Gertrude Stein and had read Ezra Pound's pronouncements about literary modernism. And all three were interested in the art work that appeared in the 1913 Armory show, the 1917 Independents' Exhibition, and in Stieglitz's galleries as well as in the private collection of Walter Arensberg, whose salons gave poets an opportunity to meet many of the artists.

The poets' perception of this international movement must be understood within a brief explanation of what modernism was thought to encompass. It began as a movement in the visual arts that rejected nineteenth-century realism. Pablo Picasso and Georges Braque began the visual style known as Cubism in 1908. Their revolutionary work rapidly spawned a number of movements, including Italian Futurism, Dadaism, Surrealism, Vorticism, and American Precisionism. As a number of art historians caution, modernism in the arts did not have one unified goal, and even apparently similar styles were often described, after the fact, by different people in different ways.[3]

Nonetheless, some generalizations can be made. Cubists, Futurists, and Vorticists all emphasized movement: their disagreement was over whether objects or artists were in motion, and how motion should be represented. The Cubists' paintings emphasized the artists' movement; to quote Gleizes and Metzinger, Cubism called attention to "the relations between the world and the thought of a man."[4] The Futurists were more interested in the fast pace of urban modernity and its effect on the perception of visual images while the Vorticists, also fascinated by movement, pictured both mental and physical motion mechanistically.[5] That is, they insisted on machine lines and they represented living forms and emotions mechanistically. Wyndham Lewis, a central figure in English Vorticism, explained that "every living form is a miraculous mechanism . . . and every . . . need produces in Nature's workshops a series of mechanical arrangements extremely suggestive and interesting for the engineer, and . . . for the artist."[6]

Between 1908 and 1912, Cubism was characterized by the use of geometrical shapes and the manipulation or fragmentation of planes and perspectives. The subject of the paintings, often human figures or still lifes, was of less importance than the artists' manipulation of forms.[7] Futurism, which began in 1909, presented a more chaotic visual field and more often took as its explicit subject the dynamic features of urban and industrial landscapes. As Robert Rosenblum suggests, the objects portrayed by Futurist art look as if they were caught in motion whereas "in Cubism the spectator, by implication, moves around static objects."[8] In 1914, the organ of English Vorticism, *Blast*, contained criticisms of the Futurists for showing "machines as moving blurs rather than as lucidly angular, cold, and impersonal objects."[9]

Other art movements also entered the scene. Tristan Tzara is usually credited with founding the movement known as Dada (as much a philosophy as a style) in Zurich on February 8, 1916, at six in the evening.[10] Dada's anti-bourgeois celebration of unreason and nihilism, accomplished with a kind of black humor, came also to inform the work of Man Ray and Marcel Duchamp, both of whom used mechanical images in their work, often as implicit comments on the dehumanizing effects of industrial and corporate modernity.[11] In 1924, André Breton founded Surrealism, which he described as "psychic automatism, by which it is intended to express . . . the real process of thought."[12] Founded as a literary movement, Surrealism inspired painters such as Max Ernst, and Breton himself described Picasso as the original Surrealist because of his juxtapositions of incongruous images and use of spatial discontinuity.[13]

Most of these movements were accompanied by published manifestoes, and this suggests that one primary characteristic of modernism in the pictorial arts and in literature was self-consciousness.[14] Allen Upward remarked in a 1911 magazine article: "It is a sign of the times that so many of us should be busy in studying the signs of the times."[15] The artists did not always follow the manifestoes. Picasso and Braque, for instance, had little to do with the later theoretical discussions of Cubism by Gleizes and Metzinger or Apollinaire. Yet the rhetoric of modernism was important in its own right. Any discussion of how artists influenced poets, therefore, must pay attention both to the language used to describe the visual arts and to the art works themselves.

Despite Gleizes and Metzinger's protest that viewers should not mistake "the bustle of the street for plastic dynamism,"[16] many modernists explicitly related their work both stylistically and imagistically to urban, technological modernity. Picabia's 1915 object-portraits, for example, were based on mail order catalogue illustrations of home appliances as well as on newspaper advertisements for appliances.[17] The American Precisionists often referred to urban experience and technological modernity, such as factories and electric signs, and they adopted the geometrical forms of French Cubism.[18] And by 1924, Fernand Léger, who was originally part of the Cubist *Section D'Or* group, praised "the geometrical austerity" of the lines of automobiles in an essay entitled "The Machine Aesthetic, The Manufactured Object, The Artisan and the Artist."[19]

Léger's later writings were influenced by his involvement in yet another movement, Purism, founded by Amédée Ozenfant and Charles-Edouard Jeanneret (better known as the architect Le Corbusier) in rebellion against the anarchy of Dada.[20] The Purists maintained an insistence on the importance of the observer in their description of a machine aesthetic and described their program as a return to Cubist principles, although they, unlike Picasso, Gleizes, or Metzinger, related the geometrical lines of both Cubism and Purism to the appearance of technological products. The name, Purism, also draws on the rhetoric of modernism, in which the terms purity, austerity, geometry, precision, and accuracy, as descriptions of a visual style, were tied to the appearance of appliances, machines, and factories.

Both visually and in various manifestoes, motion was also frequently related to machinery, to moving machine parts or to speeding cars.[21] Motion was also seen as related to human activity, that is to the artists' activity, but images of motion were more often tied to science and technology.[22] Boccioni, an Italian Futurist, celebrated "the scientific division of work in the factories . . . rapidity! precision!"[23] And Marinetti described movement in Futurist works as an effect of "wireless imagination," imaging mental activity in terms of the rapid communication made possible by the telegraph.[24] As Siegfried Giedion has argued, images such as Duchamp's superimposed and fragmented planes in *Nude Descending A Staircase* were probably related to scientific studies of motion done by photographers such as Muybridge and later used in efficiency studies.[25]

Despite the contrast between the austere machine aesthetic of Cubism or American Precisionism and the more frenzied portrayals of energy and mechanization in Futurism, it is fair to say that many modernist schools in the visual arts shared a fascination with the technological features of modernity.[26] By 1921, in his article entitled "Machinery and the Modern Style," Lewis Mumford argued that all modern styles had "the accuracy, the fine finish, and the unerring fidelity to design" of technological products.[27] The next year, Mumford offered a similar description of modernism, which he described as known for "its precision, its cleanliness, its hard illuminations."[28] By 1929, Mumford explicitly argued that modern art's precision, clean lines, economy, and accuracy were related to "the more austere forms of science and mechanics."[29]

Mumford, however, was not the first to bring modernist art to America's notice. Public awareness of these movements was introduced with the 1913 Armory Show. Readers of Stieglitz's magazine, *Camera Work*, would have known about Picasso's work even earlier, and visitors to the gallery *291*, which Stieglitz ran from 1908 through 1917, would also have been exposed to a number of American experiments in the arts, ranging from Paul Strand's and Stieglitz's photographs to the later work of the American Precisionists.[30] Although modernism in the visual arts began in Europe, by the 1920s the style that Léger labeled a machine style was associated with America.

As demonstrated in chapter 1, even without reference to a machine aesthetic, America was already associated with science and technology, loosely defined in terms of engineering feats, mechanization, and business practices as well as commercially available products. The additional association of science and technology with the visual arts, then, appears to have produced an image of modernism as an American style.

The Americanization of modernism was also directly related to the appearance of American products. Thus, Le Corbusier, whose manifesto on Purism had celebrated a crystalline or geometrical style, wrote about modern architecture by proclaiming "*American grain elevators and factories, the magnificent* FIRST-FRUITS *of the new age.*"[31] The appearance of modernity, Le Corbusier explained, had been created by the engineers and designers of telephones, baths, cars, and machines.[32] He also celebrated business and industry: "The specialized persons who make up the world of industry and business and who live, therefore, in this virile atmosphere . . . *are among the most active creators of contemporary aesthetics.*"[33]

Le Corbusier's reference to the virility of business and industry echoes the American definition of these areas as masculine pursuits and helps underline the curious situation of those defending a modernist aesthetic in America. As Le Corbusier and Mumford mention, the modernist aesthetic was related to the look of mass-produced products and buildings that were of American design. American artists and writers, then, could claim that American engineering had inspired modernism, that the aesthetic they admired was also a way of relating art to practical American reality, and, further, that modern art and poetry were masculine pursuits.

Williams, for example, drew on the suggestions of those who associated modernism with America. He probably read Mumford's "Machinery and the Modern Style," owned a copy of a book on Purist painting coauthored by Le Corbusier, and knew about modern art and architecture from his artist friends and his architect brother.[34] Williams also used the rhetoric of modernism in his comparisons between poems and mechanical objects such as cars, as well as in his emphasis on qualities such as accuracy. In 1929, for instance, he wrote that "a poem is a mechanism that has a function which is to say something as accurately . . . as possible, but . . . while we are even in the act of creating it, the words (*the parts*) are getting old and out of date just as would be the corresponding parts of a motor car."[35] In notes for a 1941 talk given at Harvard Williams says that the term *artist* itself must be made to show a relationship with *engineer* and *architect*, and notes for another article begin: "Think of the poem as . . . a machine for making bolts."[36] Marianne Moore also borrowed an emphasis on structure from modern architecture and noted that she was "interested in mechanisms, mechanics in general" (*R* 272); in 1925, she praised Strand's art, which included images of machines: "We welcome the power-house in the drawing-room when we examine his . . . perfect combining of discs."[37]

Not everyone inspired by the revolution in the visual arts was equally inspired by references to machinery. Wallace Stevens's 1918 "Metaphors of A Magnifico," like modern painting, toys with shifting perspectives: "Twenty men crossing a bridge, / Into a village. / Are twenty men crossing twenty bridges, / Into twenty villages" (*CP* 19). Stevens was interested in the visual arts; he wrote in his notebook that "the problems of poets are the problems of painters" (*OP* 160) and that poetry is "the statement of a relation between a man and the world" (*OP* 172), echoing Gleizes and Metzinger's definition of the significance of Cubist experiments with perspective. Yet unlike other modernist poets or many of the artists whom he met in New York, Stevens rarely included images of machinery in his poems or his poetics.

More often, however, poets and artists seized on the modernist admiration for American technology. After describing how Americans found art effeminate, Walter Pach added a postscript to his analysis of the place of the visual arts in America in *Civilization in the United States:* "There is . . . another phase of our subject that demands comment, if only as a point of departure for the study that will one day be given to the American art that is not yet recognized by its public or its makers as one of our main expressions. The steel bridges, the steel buildings, the newly designed machines, and utensils of all kinds we are bringing forth show an adaptation to function that is recognized as one of the great elements of art."[38]

Pach's description of an American aesthetic unrecognized by either its makers or the public reveals one problem facing modernists in America, namely that Americans did not generally find machines of aesthetic value. The problem was explicitly voiced by Henry Russell Hitchcock, Jr. Hitchcock helped introduce modern architecture to America in the Museum of Modern Art's 1932 exhibition of the works of Le Corbusier, Walter Gropius, Mies van der Rohe, and others. In the catalogue he repeated the commonplaces used to describe a modern style in painting—its concentration on technical and utilitarian factors (describing the adaptation to function also mentioned by

Pach), its clean perfection, and its debt to engineering.[39] Yet Hitchcock also complained in a 1927 article for *Hound & Horn* that America only showed good taste in areas that were not considered artistic; Hitchcock concluded that art had been usurped by the engineer and the average factory had become "more aesthetically significant than the average church and the average bathroom more beautiful than its accompanying boudoir."[40] Hitchcock's perception was shared by Bernard Smith who had read Mumford's description of American taste and who concluded in his "American Letter" for the American issue of *transition* that Lewis Mumford was "looking for beauty in America and finding it in the bathroom."[41]

A look at turn-of-the-century advertisements for bathroom fixtures in popular magazines confirms Smith's, Hitchcock's, and Pach's suspicions. Between 1870 and 1930, bathtubs became common household fixtures as manufacturers slowly perfected techniques for the mass production of steel-clad and enameled tubs.[42] The clean lines, adaptation to function, and geometric shapes characteristic of a machine aesthetic are found in American bathroom fixtures. Manufacturers downplayed the appearance of their products, however, in order to sell them to the public. An 1896 advertisement from the midsummer issue of *Muncies Magazine* emphasizes that a steel-clad bathtub is graceful by using a drawing of a tub on a pedestal in the center of the temple of Zeus at Aegina. The copy reads "The Temple of Cleanliness." Both the image and the copy relate cleanliness to godliness, and hide—rather than celebrate—the tub's lack of ornamentation.

Artists were then forced to acknowledge that while the appearance of technological products was arguably American, the taste for technological lines was not.[43] Modernist poets faced a similar difficulty when they adopted the rhetoric used by the visual artists. Williams explained, for example, that poetry "bare, stript down, has come to resemble modern architecture."[44] This reference to architecture involved his competition with his brother, Edgar, who was a prize-winning architect. Sibling rivalry aside, his brother's work made Williams aware that architects could design structures that were, in the rhetoric of international modernism, both modernist and American. Le Corbusier compared engineers, architects, and the builders of American business.[45] He also compared houses with machines just as, by 1944, Williams compared poems with machines.[46] In that a machine is a combination of resistant parts arranged to do work, a poem may aptly be described as a machine. Given the American context, however, there were a number of problems that arose from comparisons between poems and machines or modern buildings. As I showed in chapter 1, the American public appeared to value literal machines, buildings, and factories, but they did not respond well to imagist or other innovative poetry.

Modernist writers followed the artists in others ways that proved problematic as well. Drawing in part on the celebration of a machine aesthetic in art movements such as Purism or American Precisionism, writers at times described poetry, nature, the mind, and the self mechanistically. In 1939 Stevens called the ocean a "universal machine" (*OP* 82).[47] In R. M. Thompson's "Genuflections to the Engine," the speaker moves "machine-like, precisely, exactly."[48] And Williams stated that "knowledge itself is . . . a machine" (*EK* 63). Williams also wrote to James Laughlin that the "mind is a queer mechanical machine" and he compared people to "Pasteur's crystals" (*EK* 26), thereby linking the images of crystals used by the Purists to describe their visual aesthetic with Pasteur's scientific work on crystals.[49] In these mechanistic or crystalline descriptions Stevens, Thompson, and Williams were in part following the visual artists' description of a new, geometrical style. Indeed, whatever the public's response, the use of mechanistic images in poetry and art did distinguish modernist work from the more sentimental and clichéd nineteenth-century styles against which the artists and writers rebelled.

At the same time, mechanistic descriptions had long been commonplace outside the world of literature. Medical textbooks, health pamphlets, magazine articles, and advertisements frequently discussed the human body as a machine or described food as "fuel for the human engine."[50] W. C. Blum's 1921 "American Letter" explicitly compares Williams's writing to that found in Dr. William Osler's medical textbooks.[51] Metaphorical images of mechanisms or machines thus marked both a modernist and an American sensibility, and such metaphors in one way filled Oppenheim's and Brooks's prescriptions for an art that acknowledged practical American reality. Mechanistic images could be seen as attempts to provide a language for technology, and so to humanize it, yet they were also easily read as technologizing humanity; that is, Thompson's image of a machine-like person seemingly glorifies just those effects of industrialization and business efficiency that Brooks and Dewey protest.

The ease with which mechanistic references can be misunderstood is seen by examining Le Corbusier's celebration of engineers, architects, industrialists, and businessmen. Le Corbusier praised America for its development of machines both as aesthetic objects and as tools, and he saw that mass production techniques were capable of improving people's lives. Yet he also objected to assembly lines and warned that the "machinery of Society, profoundly *out of gear*, oscillate[d] between an amelioration . . . and a catastrophe."[52] Although he distinguished between better and worse uses of technology, Le Corbusier did not seem to realize that mass production was made possible by assembly lines and that part of the social unrest he described could be related to the conception of individuals and social institutions as machines. The dual nature of mechanization was more obvious to many American modernists; not only were both the gains and the ill effects of industrialization and corporate practices more evident in the daily lives of most Americans, but the writings of men such as Brooks and Bourne had also made readers aware of the problems attending modernization.

Another problem also arose from the poets' adoption of the artists' language. Artists could look to the lines and hard-edged qualities of industrial structures or technological products, as is apparent in the geometrical styles of Cubism and Purism. It was less clear how poetry or any writing might embody a machine aesthetic, even though Futurism, Dadaism, and Surrealism included literary programs, which emphasized stream-of-consciousness narratives and irrational or startling juxtapositions of images and voices.

English writers, also inspired by developments in the visual arts, faced the problem first. In 1924, Virginia Woolf described the advent of modernism as a revolution in sensibility by declaring that "on or about December 1910 human nature changed"; as has been pointed out, for Woolf human nature changed about the same time that the first Post-Impressionist Art Exhibition opened in London.[53] How exactly poets might follow the artists was, however, unclear at first.

T. E. Hulme was among the first in England to suggest that a modern style should be geometric, hard-edged, and mechanical. Hulme wrote that modern poetry, like modern art, would have to be "austere, mechanical, clear cut, and bare" even before he found a poetic style that might call forth such adjectives.[54] Hulme denied that his new aesthetic borrowed from science or technology, saying that an artist does not use "mechanical lines because he lives in an environment of machinery," yet he also noted that art's "association with the idea of machinery takes away any kind of dilettante character from the [modern] movement and makes it seem more solid and more inevitable."[55] In another early essay, he wrote that philosophy, "tempted by science, fell and became respectable"; the resulting scientific view of the world, he continued, made a society in which the "days of adventure were gone. . . . Here was no place for the artist."[56] Hulme's description of poetry, with a vocabulary borrowed from the respectable domains of science and technology, most obviously is an attempt to rediscover a place for art in the modern world. And the alacrity with which poets such as H. D., Pound, Fletcher, and others adopted styles that fit his description—precise, impersonal, bare, mechanistic—suggests that there existed a climate in which Hulme's strategy for the defense of poetry made sense.

Ezra Pound's early poetry reveals that he, like Hulme, wanted to revolutionize poetry before he knew how to do it. He felt that poetry should break with the tradition of English Georgian poetry and American genteel poetry, but at first he was not certain what form a new poetry would take.[57] Thus, he described his 1908 *A Lume Spento* to Williams as beginning something revolutionary: "I, of course, am only at the first quarter-post in a marathon."[58] Later, Pound himself dismissed the volume as a "collection of stale creampuffs."[59] By 1912, however, Pound had evolved a description of his new poetry, based in part on the poetic practice of Hulme and H. D., but also repeating the language of manifestoes in the plastic arts, which had been available on the continent since 1909. Pound's Imagist Manifesto, written in 1912 as a note to accompany some of Hulme's poems and first published in America in 1913, advised writers to use "no superfluous word, no adjective which does not reveal something, . . . either no ornament or good ornament," and to go "in fear of abstractions."[60]

Pound did not frequently refer to machines, but he did argue that the "serious artist is scientific in that he presents the image of his desire . . . precisely."[61] Inspired by numerous sources, including Ernest Fenellosa's work on Chinese characters, Pound's conception of scientific objectivity does not wholly parallel the visual artists' understanding of science. Yet the language in which Pound described Imagism to America did accord with the language in which a machine style was being described.

The year that Pound's instructions on how to write Imagist poetry appeared in *Poetry* was the same year (1913) that the Armory Show introduced the America public to modern art. Within a decade, Williams's poems as well as the poetry of Moore, Stevens, and Amy Lowell were widely, if at times misleadingly, associated both with modern painting and with Pound's Imagist movement.[62] Williams's retrospective description of his interest in modernism acknowledges the influence of Pound's manifesto as well as of experiments in the visual arts. Williams explained that the writers he knew were "closely allied with the painters" and "followed Pound's instructions" (A 148).

American modernists used Pound's 1913 essay for *Poetry* in their adoption of a bare, stripped down literary style that allied itself with modern art and architecture, yet they also defined a modern poetic style in distinctive ways. Moore and Williams, especially, followed Pound's instructions to avoid unnecessary words or abstractions, but they further paid attention to the look of poems on the page.[63] Williams's 1937 "Classic Scene" (*CEP* 407) provides a good example. Although there is argument over the exact nature of the relationship between Williams's poem and Sheeler's 1931 painting *Classic Landscape*,[64] it is generally agreed that the aesthetic as well as the title of Williams's poem owe a debt to American Precisionist painting. Meyer Schapiro describes the Precisionists' work, which is related to Cubism, as asserting a radical empiricism in the artists' photographic realism, by which viewers were invited to see the beauty, rather than the utility, of American barns, factories, and other structures.[65] Williams's "Classic Scene" similarly looks to the buildings of industrial America for both its subject and its style.

In regular quatrains, "Classic Scene" describes a powerhouse and two metal stacks "commanding an area / of squalid shacks." One stack is smoking; one, passive. No judgments are voiced and only one word, squalid, is explicitly evaluative. The irony in the title and the image of the smokestacks as a couple sitting on a single chair humanize the scene: the smokestacks are as familiar as our grandparents, as Henry Sayre remarks.[66] It may be that Williams thus ameliorates the industrial landscape, as Sayre also argues.[67] Yet if industry is domesticated by Williams's metaphor, the domestic is thereby technologized; the implication of both the metaphor and the inclusion of the squalid shacks dominated by industrial images is that industrialization has changed American life. That is, just as the industrial complex dominates the visual field described, so the industrialists command the social arena. There is a tacit critique of modern life in the poem at the same time that the appearance of the landscape is admired.

Williams's style, like Sheeler's, celebrates the aesthetic of American industrialization. He takes a machine aesthetic, drawn from not only the painters but also the actual appearance of the modern American landscape, and humanizes it with the image of the smokestacks as figures sitting in chairs. Yet the style of the poem allies it with the factory, not the shacks. Like the factory, the poem is a carefully structured design with its four quatrains and single concluding line. As Sayre concludes, form and content are at odds.[68] By 1937, when he wrote "Classic Scene," Williams was well aware of the tension in his poem. The use of a machine aesthetic in poetry, however, presented potential problems that most people in the 1910s and 1920s could not easily articulate, let alone control or address.

That Williams, Moore, and Stevens became self-conscious about the tensions involved in the use of mechanistic metaphors and of a machine aesthetic may stem from the various ways in which they understood machines and American modernity. Modern artists whose work the poets admired celebrated American technology and associated it with structures such as factories, mass produced products, and even at times with business. Moreover, these three poets had a healthy respect for the extraliterary uses of machines and other American products from medicines to typewriters. Moore and Williams, especially, believed that poets might use technology both literally and to define an aesthetic that would be more relevant to their age than the aesthetic of nineteenth-century poetry. Santayana, Dewey, and others, moreover, had implied that art and poetry needed to be reconnected with the reality of technological America.

On the other hand, all three poets deplored how American advertising fostered an image of the products of technology as the cure for all America's ills. Even as they found aspects of American technology attractive, each of the poets felt that technological modernity had lowered the esteem in which poets and poetry were held and that poetry needed to be defended against more common and less attractive American values. They were also aware that European audiences, even those who labeled the style of modernist art and poetry "American," thought that artistic creativity was stifled in America.

Pound, for example, originally announced that Imagism, a movement founded on English soil, marked the beginnings of an American Renaissance; in 1912, he wrote to Harriet Monroe and predicted an "American Risorgimento . . . [t]hat . . . will make the Italian Renaissance look like a tempest in a teapot!"[69] Yet when Pound visited America, he "found no writer and but one reviewer who had any worthy conception of poetry."[70] In short, Pound was skeptical that America really could produce strong poetry. Others not only shared his skepticism but believed that America's commercialized technology showed the sole form that American ingenuity could take. Lewis Mumford celebrated the aesthetic of American technological products in his 1929 *American Taste,* but in 1925 he wrote to Stieglitz after lecturing on American culture in Geneva that Europeans were "very unwilling to admit anything can come out of America but ford [*sic*] cars and fountain pens."[71] Similarly, most of the 1928 American issue of *transition* echoed Marcel Brion's response to an "Inquiry Among European Writers Into the Spirit of America," where the country that discovered the "beauty of an ice-box" was sarcastically said to be "really above literature."[72]

American poets who stayed in this country, then, tried to address two audiences: They wished to prove to the world that America could produce great art as well as great automobiles and they wanted to reach an American audience that seemed to care little about art of any kind. This complex of attitudes towards poetry, technology, and American modernity informed the poets' understanding of how comparisons between poems and machines worked and did not work in an American context. Comparing poems with automobiles might call attention to America's creativity, and it might associate poems with products that had captured the public's imagination. Yet not being material objects with obvious uses, poems were not easily defended to practical American consumers. Indeed, to defend the utility or practicality of poetry was to measure poetry by the same standards used to assess vacuum cleaners or medical technology.[73] As Kenneth Burke wrote, "One cannot advocate art as a cure for toothache without disclosing the superiority of dentistry."[74]

Moore's analogies between technology and poetry often avoid comparing poems with physical objects and instead ally the ingenuity or mental speed of inventors with the ingenuity of writers who also show "a scientifically potent energy."[75] Yet, like Williams, Moore praised the products of American technology for the ways in which they had improved the quality of life in America. "Equipment," as she put it, "is not invariably a part of culture"; however, American women have been "[a]ssisted by the typewriter, the sewing-machine, and the telephone."[76] At times, Moore also emphasized the usefulness of poetry, as when, in a 1956 talk, she compared poetic technique with technology, pointing out the Greek root, "*tekto*: to produce or bring forth—as art, especially the useful arts" (*R* 173). In her poem, "Poetry," she also invited a comparison between the utility of poems and the utility of commodities marketed as *genuine*. In so doing, Moore set out to educate her audience. Although she did not claim that poetry could cure toothaches (to use Burke's example), she worried that poetry could not convincingly be defended as useful to an American audience.

Moore's assessment of her American audience appears to have been accurate. The Lynds' 1929 study documented that Americans were very interested in the useful arts and did not count poetry as such an art. The Muncie library, for example, found that books on what the library categorized as the *useful arts*, including technology, advertising, and salesmanship, were in such demand that they could not be kept on the shelves.[77] Between 1903 and 1923, reading of library books on the useful arts increased sixty-two-fold. The library apparently did not categorize poetry separately; however, readership of the fine arts and of literary fiction increased less than half as much as readership of books on practical, useful subjects.[78] In such a context, how could poets defend poetry as a useful art?

Nonetheless, writers, following the artists, still attempted to borrow the prestige of technology in their defenses of poetry. To do so, they exploited the way technological products, scientific discoveries, scientists, and engineers were often confused in common usage. Hygiene was one area that was frequently used.[79] When Le Corbusier mentions an American machine aesthetic, he refers to both baths and hygiene in one sentence, conflating appliances and applied science. Similarly, Mumford invokes the appearance of bathroom fixtures and the techniques of mechanical mass-production when he praises the accuracy, precision, and fidelity to design of a machine style. The poets, in effect, exploited the way hygiene could be understood to refer to medical advances, to the products of technology, and to technological design.

Most often, and perhaps most simply, writers associated *cleanliness* with the short lines and the avoidance of adjectives Pound had recommended. Certainly, these stylistic features might be called clean, but there is also a pun involved in equating the cleanliness of hygiene and the cleanliness of a style. American advertisers, for example, made no such equation between hygiene and a lack of ornament. The appearance of hygienic fixtures (which artists celebrated), the improvements in health care, and the scientific discoveries on which they rested were easily conflated, however. Bathrooms have clean lines; they promote cleanliness and health, an obvious good; and they result from laboratory work, generally conceived of as requiring antiseptic conditions. When Williams described a "cleansing of the 'word'" (*EK* 6) in the late 1920s or when, in 1917, Leo Stein recommended that the arts provide a "soul's hygiene" both invoked the prestige of American hygiene and technology for the arts and at the same time associated themselves with the aesthetic of international modernism.[80]

Borrowing the vocabulary of artists when American advertisers used the same vocabulary in different ways posed problems for the defense of modernist poetry. And inviting readers to associate poetry with more tangible and more commercially successful achievements did not usually lead to an admiration for poetry. Both Moore and Williams were self-conscious about the conflict between their desire to claim the virtues and values of technological modernity and their recognition that, within America, poetry's value might need to be defended in some other way.

Like Moore, Williams shifted ground on the question of whether American poetry could be defended by analogy with more practical disciplines. In 1947, he wrote to Burke that his own enterprise was "little different from the practical deductions of an Edison" (*SL* 257). Williams draws on the image of Edison as an amateur inventor whose primary strength was native intelligence and he draws on the accepted practicality and usefulness of Edison's inventions. He agreed, nevertheless, with the import of Burke's 1931 suggestion that poetry might serve the need "to 'corrode' the practical."[81] In the late 1920s, before he could have read Burke's suggestion, Williams wrote that when science or philosophy "seek ends . . . they deem 'practical' . . . they become superstitions" (*EK* 92). Williams's rhetoric works in several ways. His references to the practicality of poets as comparable to the practicality of Edison draws on the popular prestige of American inventors even as his description of scientific or philosophical "practicality" (the quotation marks are Williams's) redefines what is practical.

Again, although Williams's intention was both to reach and to educate his readers, he was aware of the difficulties of defending poetry as practical to an American audience. Williams's uneasiness is particularly evident in a 1954 address he gave to executives of Bell Telephone, that is, to representatives of technological and corporate America. In his address, Williams compares poets to engineers, an analogy he first made in a 1921 article for *Contact*. The comparison served, in part, to defend the importance of poetry by an appeal to the widely recognized importance of engineers. As Wallace Stevens noted, the "prestige of the poet is part of the prestige of poetry" (*OP* 174). Williams specifically used his analogy to argue that poets and engineers shared both an attention to design and a kind of practical know-how; to the Bell executives, he explained that "fundamentally the engineer and the artist agree."[82] Williams also acknowledged how his audience actually responded to poetry: "The poem is not attractive, as a general rule to men of affairs. . . . rightly so, since for the most part, it has nothing to communicate to them."[83]

Although addressing a different audience (and one not seated before him), Williams's 1921 essay works similarly. The essay argues primarily with those such as Pound, Hartley, and Mina Loy, who had given up on America:

> It has been by paying naked attention first to the thing itself that American plumbing . . . indexing systems, . . . and a thousand other things have become notable in the world. Yet we are timid in believing that in the arts discovery and invention will take the same course. And there is no reason why they should unless our writers have the inventive intelligence of our engineers and cobblers. (*SE* 35)

Williams again redefines the virtues of technologists and inventors, emphasizing creative intelligence. The list of "things . . . notable in the world" refers, however, to technological products and to the international reputation of such products, that is to their fame. The fame of American plumbing and indexing systems, in fact, involved both commercial success and tangible results.

Williams, like Moore, genuinely admired American inventions. He also invoked the prestige of American technology in order to call attention to practical intelligence as the true American strength, a strength that might produce American poetry. Those who had left the United States, however, believed that the love of commercial products made America a country that could not nurture the arts. Moreover, given popular American definitions of what made engineers successful, invoking achievements in fields where results were more tangible and more lucrative was problematic: How was the value of poetry to be defended when the criteria of value (usefulness or profit, for instance) in the field to which poetry compared itself could not easily be claimed for poetry? Hart Crane's mistrust of poetic values borrowed from technology stemmed from his acknowledgment of just this problem: "When you ask for exact factual data . . . etc., from poetry—you . . . ask its subordination to science."[84]

Williams believed that the news poetry could deliver was valuable, but both he and Moore uncovered the difficulty of describing poetry's value to an American audience given that the vocabulary of modernism invited comparisons between poetry and fields that had more prestige and produced more literal results. Moreover, the question of what value poetry had was one the poets shared with their American audience. Williams wrote to Moore in 1944: "As you yourself said once, in substance, so well: what good is all this stuff? It doesn't work, it doesn't split any rocks after all. . . . we all expect our verses . . . to do work, to make a better world of it. But they are so weak" (*SL* 231).

Both Moore and Williams, at times, offered an alternate view of what technology revealed about the American character. This view tried to avoid the comparison between the tangible or profitable results of technology and poetry. Moore, for example, asked whether "the poet and scientist [did] not work analogously?" (*R* 273). Williams concentrated on the inventive intelligence of engineers. He also defined science in a way that recalled Gleizes and Metzinger's definition of Cubism when he wrote that science "changes nothing, invents nothing . . . it [only] brings its material into a certain relationship with the intelligence" (*EK* 129). Thus practicality and utility, the very traits that Williams at times appropriates for poetry, are denied to science, which is redefined as primarily theoretical by separating the strengths of science and engineering from their tangible results or from the invention of products.

In a similar manner, Williams caps an argument on the superiority of art over science with a quotation from Pasteur, and he includes Pasteur in the company of poets because he "played around" with non-standard material (stale beer) and worked without an eye to results: He "damn near starved. . . . Why not welcome him?" (*EK* 118). Williams here moves from an identification of the products of technology and art to a comparison between pure scientists and artists; that is, poetry does not so much look like technology as arise from the same kind of creative inquiry. In short, it is not the product but the processes used by scientists, technologists, engineers, and poets that are comparable. Pasteur's method of working without an eye to results is compared with poetic experimentation.

Williams makes the same point when he distinguishes between pure and applied science:

> There is a choice . . . in the arts. . . . It is the same that exists in chemistry, life at large or whatever it may be. It is the "pure" as opposed to the "practical." Pure chemistry as opposed to industrial chemistry. . . . applied design at one pole and—the artist. . . .
> . . . Surely there should be a place . . . for [pure writers] if

> not in the practical world. It is the lack of this primary consideration which is offensive in a place like the United States. (*EK* 116–17)

It is in this section that Pasteur is mentioned as one who played around, and thus "advanced 'pure' bacteriology" (*EK* 118). The claim is clear: Pure science is not practical but creative, just as poetry is; therefore, poets, poetry, scientists, and experimental science should all have a place in America.

Moore and Williams were not alone in identifying the creative spirit behind scientific and technological products. Examples are numerous. Robert H. Lowie's chapter on science in *Civilization in the United States*, for instance, argues that scientists must be creative, and O. S. Beyer, Jr., writing on engineering in the same volume, claims for his discipline service (that is, utility), efficiency, and creativity.[85] Similarly, Enrico Prampolini wrote of the machine as a common symbol "of the mystery of human creation."[86] Hart Crane not only admired Roebling's bridge, but seriously considered writing his biography.[87] The idea that both scientists and engineers were creative was not limited to an avant-garde or even to a relatively highbrow audience; the hero of an early 1930s movie, *I Am a Fugitive from a Chain Gang*, returns home from the Great War saying he wants "to get away from routine . . . and create" by becoming an engineer.[88]

The suggestion that art and science have a common ground is appealing, yet it seemed insufficient for the American modernists simply to note the creativity behind both art and science. Even the notion that writers exercised creativity in a purer form, by avoiding the taint of technology, did not adequately defend poetry's value in the American context. Since the success of technology and science did not seem to rest most immediately on the pure imagination of scientists but on tangible and commercially measurable success, poets could not gain prestige for poetry, as they wanted to, by emphasizing art's or science's creativity.

Thus, even as they focused on the risk taking and experimentalism of science, the poets had to concede that America did not usually value what was impractical. The attempt to redefine the reason for Pasteur's fame foundered on the recognition that he was popularly known and esteemed because of the practical, and highly visible, results of his work as applied by Lister and by those who marketed baby formulas and bathtubs. That is, the American public did not value science primarily for its pure creativity. This is tacitly acknowledged in the way Williams appropriates features of technological products—accuracy, effectiveness, practicality, cleanliness—and applies them first to the scientific process and then to poetic creation and poems. The metonymic shift focuses on the process not the products of science so that poetry, insofar as it is metaphorically associated with science and technology, can share some of the high regard popularly enjoyed by technology. This requires that technological science be redefined as pure science. And, early views of Einstein aside, the traditional status of pure science in America was not what poets wished to invoke.

The poets were often self-conscious about the ways in which they tried to exploit the looseness of the words science and technology in common usage. In his attack on science and technology in the late 1920s, Williams made "a special plea for the usefulness of the field of art" (*EK* 125), which he contrasted with scientific utilitarianism. Of the accumulations of Science, he said "An Indian knew more of fire than the designer of the Bessemer smelter. . . . That it [science] is valuable, useful, we know. This is unfortunate, since apart from that it is worth nothing" (*EK* 80–81). In short, he resisted the popular definitions of science as useful and valuable, even as he insisted that poetry was of both use and value in a different way. At the same time Williams called popular estimates of science into question, however, he also appealed to the popular respect for utility when he argued that the field of art was useful. Thus, art is accorded just those features denied to science, features that historically do seem to have been why science (or, rather, technology) was valued by the public.

Even for those who admired modernism, the strategy of using and redefining public rhetoric was not always effective. In 1925, for example, despite his early call for a literary Pasteur to revitalize the arts by providing a *soul's hygiene*, Leo Stein protested that art could not simply appropriate adjectives, to be used metaphorically, from the sciences: "Accuracy could hardly be more utterly in default than in a precise application of the word 'accurate' to literature."[89] Moore responded to Stein that despite his accusation that writers attempted "'a pseudo-hardness . . . which . . . is really singularly inexpressive'," there was, indeed, such a thing as accuracy in literature, which involved "our manifold ferocities and ungainly graces, . . . a corollary to momentum."[90] Moore often identified accuracy with motion; as one of her poems says: "We are precisionists, / not . . . arrested in action" (*Comp* 59). Drawing on Cubist and American Precisionist definitions of motion, Moore generally suggests that it is the motions of the mind of which art gives a scientifically accurate portrayal. Williams says that art "measures, is accurate and devotes itself to the facts" (*EK* 93), and Stevens proclaims that the "final poem will be the poem of fact in the language of fact" (*OP* 164). All similarly attempt to equate poetic and scientific accuracy, although the poets differ in where they locate poetry's accuracy.

In defining poetry as useful, precise, or accurate, the poets found themselves accused of being merely metaphorical and still of leaving poetry with no clearly defined place in a technological society. The difficulty of finding a favorable description of poetry's relationship to a technological society is in part responsible for the ways in which the poets alternately discuss poetry in technological language and technology in terms of the ingenuity or creative methods that produced it. Neither strategy seems finally to have solved the practical problems facing poetry. The poets did not thereby gain an audience for their poetry, and they themselves were sometimes uneasy about how practical and technological America could be included in, let alone be helped by, poetry.

Moore, Williams, and Stevens, however, were self-consious about how art, poetry, and American advertisements for technological products all used adjectives metaphorically. Although he generally avoided the use of technological language or imagery in his poems, Stevens stated the proposition most dramatically: "Usage is everything" (*OP* 159). Ultimately, recognizing that America's public language was the part of American reality that poetry could include and revalue was one of the poets' triumphs.

3

Science and American Modernism: Saint Francis Einstein

MANY of the writers whose interest in technology is documented in the preceding chapters were also following at least popular press accounts of the breakthroughs in pure science, especially in physics, and they eventually drew on what they understood of Einstein's, Whitehead's, and Heisenberg's work to describe what might be called a poetry of process. Most writers did not confront the tensions implicit in their descriptions of poems as alternately objects and process; many were probably not aware that they generally looked to technology for their vocabulary and analogies when describing poems as objects and to the new physics when describing poems as actions or process. Nonetheless, it is useful to isolate for consideration the descriptions of poetry as process, the analogies used to present poetry as process, and the uses such analogies served in the first half of the twentieth century in America.

The image of poetry as process or as energy in motion was not new in twentieth-century America. Hart Crane, for instance, seems to follow Henry Adams in discussing *energy* and *combustion* together. Presenting a modern myth, which is articulated with some ambivalence in "Cape Hatteras," Crane writes: "Man hears himself an engine in a cloud."[1] He also refers to Walt Whitman, revealing a nineteenth-century source for the idea that mechanical energy could be an analogue for divine spirit.[2]

The insistence on energy and change was a commonplace of the avant-garde art world by the first decade of this century. Demuth's poem on Duchamp sounded a familiar note, suggesting that most artists "stop or get a style. / When they stop they make / a convention. / That is their end."[3] The appeal to creative process is clear, as it is in Stevens's chant: "Nothing is final . . . No man shall see the end" (*CP* 150). Stevens, like Crane, associates process with Walt Whitman. Not surprisingly, the style seen as appropriate for a poetics of motion and energy was often Whitmanian—long lined, at times with lists of objects held together in an interior monologue. Williams's style in "The Wanderer," for example, evokes Whitman as well as a poetic energy to match the urban power and restlessness depicted:

> The women's wrists, the men's arms red
> Used to heat and cold, to toss quartered beeves
> And barrels, and milk-cans, and crates of fruit!
> . . . .
> . . . Everywhere the electric!
>
> [*CEP* 7]

From the 1920s through the 1950s, Einsteinian physics was often enlisted as a sanction for this poetics of process. In 1955, referring to his "relativistic or variable foot" (*SL* 335), Williams wrote that the structure of the poetic line "is where aesthetics is mated with physics" (*SL* 330). Later he added: "When Einstein promulgated the theory of relativity he could not have foreseen . . . for a certainty its influence on the writing of poetry" (*SL* 335–36). As early as 1931, in *Axel's Castle,* Edmund Wilson discussed Proust's affinities with modern physics insofar as Proust believed that "all our observations . . . are relative" (p. 157). Wilson also noted that like "Proust's or Whitehead's or Einstein's world, Joyce's world is always changing" (p. 221). Change and motion, seen as part of a relativistic universe, were the catch phrases for which Einstein's theories seemed to offer support.

Bergson's philosophy was also associated with Einstein's discoveries. Indeed, Wyndham Lewis, in his 1928 *Time and Western Man,* found Einsteinian physics responsible for overturning a pre-War attention to formalism and reviving "Bergson's psychological time-conception, and his doctrine of a 'creative' flux—the physiological, organic, view of nature."[4] Lewis may have a point in his account of the sources of Bergson's popularity, but appeals to the principles of dynamism, motion, and process were new only insofar as they now were said to be founded on the insights of Einstein and Bergson. By extension any style, such as Whitman's, that stressed motion was soon related to Einstein or to Bergson. Thus, in 1927, reviewing Virgil Jordan's *Yale Review* article on the American economy, an editor for *Hound & Horn* judged: "The American 'secret' is not mechanization but mobility of production. . . . The idea is an interesting application of Bergsonian philosophy to economics. And Mr. Jordan achieves a movement of style that satisfies a Bergsonian critique of expression."[5] If earlier in the century science was identified with its practical technological offshoots, after Einstein, technology itself was sometimes redefined in light of the vocabulary of theoretical science. In *Time and Western Man*, for example, Lewis includes a diatribe against Einsteinian philosophies of time and motion, which he says have been connected with Bergson "and the motor-car."[6]

With no sense of contradiction, science and technology were simultaneously invoked to provide images of modernity. In 1928, Walter Lowenfels described the modern age as the *Machine or Einstein Age* in an article for the magazine, *transition*.[7] Similarly, Williams's original plan for his *Collected Poems* included six sections, the last two of which were to be called "The Poem as a Machine" and "A New Way of Measuring," with measure related to Williams's understanding that Einstein had dictated new forms for poetry.[8] Form "is the physics of verse," Williams wrote in a 1948 essay, adding that it "would be a stupid insult to Einstein to write him a sonnet."[9]

Although in Williams's section titles and Lowenfels's article there is little sense that the age needed to choose between machines and physics in seeking an emblem of the times, the models offered poetry, as Williams suggests, were both stylistically and conceptually different. Most writers drew on both technology and pure science, sometimes in the same breath, but there was also a tension between the two types of poetry that the different analogies were seen as supporting. In order to explore this tension, it is necessary first to consider in more detail the ways in which the analogy between physics and poetry was used, especially in America.

In many ways, the warm welcome that newspapers gave to Einstein in 1921 when he visited America and the physicist's continued early popularity in this country are not easily explained. Whether one looks at late nineteenth-century American journals such as *Popular Science Monthly* or at the 1922 chapter on science in Stearns's *Civilization in the United States*, one finds science defending itself against literature and technology, its own claims to being the touchstone of truth and civilization at least as edgy as the literary claims made in the face of science and technology.[10] Scientists were not, in general, seen as heroes in the United States. Thomas Edison, described as a practical man, was a folk hero, but the name of A. A. Michelson, the first American Nobel Prize winner in physics (in 1907) and the man whose work paved the way for the relativity theory, was hardly known by the lay public. Similarly, the American scientist Josiah Willard Gibbs, who laid the theoretical foundations for modern physical chemistry, was relatively unrecognized in his own country.[11] The public image of nineteenth-century science varied in part depending on whether the reference was to the more theoretical sciences or to the more empirical natural sciences.[12] Whatever the science involved, the stereotype of the scientist remained that of the impractical, often upper-class, man, hardly the stuff of which the American ideal was formed.[13]

Even more than Pasteur, however, and without the association between science and useful results, Einstein and the Curies captured the popular imagination of America. Wyndham Lewis expressed his feeling that within "a few years of the arrival of Einstein upon the european scene, the layman . . . knows more about Relativity physics than any layman has ever known about the newtonian cosmology."[14] The reception was at least as warm in America, where Einsteinian physics filled the popular press between 1919 and 1922. By 1921, the *New Republic* was editorializing about Einstein as a fad on a par with chow dogs or alligator pears; with the physicist's visit to America in 1921, he was more or less adopted as a native son.[15]

Writers were particularly interested in Einstein. The cover of the January 1921 issue of Williams's magazine, *Contact*, defined contact in terms of energy, a clear reference to Einstein's redefinition of reality as energy or process. According to Williams, contact (the magazine and the word) was "a vast discharge of energy forced by the impact of experience into form."[16] A November 1921 issue of *Broom*, edited by

Kreymborg and Loeb, contained a lengthy parody of an anthology of poems as they would have been written by various authors after an exposure to the new physics.[17] By 1927, Hart Crane could describe a new boss as a *cultured man* because he discussed "Aristotle, Einstein, T. S. Eliot, Gertrude Stein, etc."[18] The same year, for those who did not have access to foreign periodicals, *Hound & Horn* reviewed an article by Marinetti from *transition* on "Futurist Standards of Measurement," and the following year included a review by R. P. Blackmur of Lewis's anti-Einstein diatribe, *Time and Western Man.*[19]

At times, the literary world's adoption of Einstein was simply a way of granting authority to poetry. *Camera Work's* 1909 defense of *straight* photography used a similar argument, conflating the camera and the sort of image produced, when it suggested that to say that photography is "scientific is at once the significance and the measure of its value."[20] That which was scientific was seen as having a concrete, objective status. Thus, in 1937, Williams claimed that Marsden Hartley should be less subjective and "more of a scientist" (*SL* 168). Even Williams's defensive proclamation that art "alone remains always concrete, objective" (*EK* 56) indicates his respect for the traits most often associated with science—its objectivity, its access to truths about the concrete world.[21]

As with technology, the truth value of science was appropriated by the artists who justified their practice and style by appeal to Einsteinian physics. Hart Crane, though more sarcastically than many, noted of truth's name: "Her latest one, of course, is 'relativity'."[22] Williams's and Wilson's appeals to the truth value of science are less mixed. Wilson elevates literary achievements by saying that writers are "working, like modern scientific theory, toward a totally new conception of reality" (p. 297), and Williams invokes not only Einstein's creativity (*SL* 252) but also scientific *hardness* generally to underwrite poetry, justifying his search for a new measure by comparing the object of his quest with a new element "predicated by a blank in the table of atomic weights" (*SL* 243). The point of such analogies was to argue that poetic experimentation was as firmly connected with verifiable discoveries about the real world as scientific experimentation.

At the same time, science as a pure discipline seemed more easily compared with poetry than technological science. At least until the

1940s, theoretical science seemed to have virtues analogous to the virtues of poetry, with none of technology's more tangible and less easily matched products. And the new physics seemed to offer a way to defend poetry. Hart Crane's assertion that all science is "exact knowledge and . . . in perfect antithesis to poetry" reveals a common stereotype, but does not take into account how others paradoxically used the hardness and exactness of science to vindicate the apparent subjectivity of poetry.[23] For one thing, the Curies' discoveries suggested that scientists held the material world less firmly in hand than had been suspected. For another, the association of Bergson with Einstein suggested that the subjectivity with which poetry seemed to be concerned was scientifically important. As Williams put it, "mathematics comes to the rescue of the arts" (*SE* 340).

Obviously, the rescue mission involved strained logic. To give another example, Williams used physics to certify that modern poetry was tied to reality: "It may seem presumptive to state that such an apparently minor activity as a movement in verse construction could be an indication of Einstein's discoveries . . . but such are the facts. . . . The verse I can envisage . . . comes much closer in its construction to modern concepts of reality."[24] In his 1935 essay, "A Poet That Matters" (perhaps punning on the new ideas about physical matter), Stevens pointed out the irony of "sticking to the facts in a world in which there are no facts" (*OP* 254). The irony is clear: Physics did away with solid matters of fact, and thus gave poetic knowledge a new respectability; we know this because science, which we respect, tells us it is a fact. To put it another way, the objectivity of science was called on to validate the apparent subjectivity with which poetry seemed to be concerned.

Many references to the new physics were quite loose, ranging from Frost's remark, made in the 1930s, that science "put it into our heads that there must be new ways to be new,"[25] to Henry Russell Hitchcock's 1927 article about the decay of American civilization, where he casually asserts that speed "even in connection with the movement of civilizations is wholly relative."[26] Not only physics, but the sciences generally were seen as experimental, and their experiments were seen as analogous to experiments in the arts. In a 1913 piece in *Camera Work*, De Zayas observed that art "proceeds toward the unknown, and that unknown is objectivity. It wants to know the essence of things . . . following the method of experimentalism set by science."[27] While De Zayas's reference is particularly to empirical science, the image of experimentalism was also applied to Einsteinian physics. Marianne Moore's 1932 poem, "The Student," quotes a *New York Times* article on Einstein, having the scientist answer the question "When will / your experiment be finished?" by saying, "Science / is never finished" (*Comp* 101, 278).

The larger context of Moore's poem, which is primarily on American ideals and institutions, suggests as well how the image of scientific experimentation was loosely associated with political experiments, and particularly with American democracy. The free play of the mind to which Einstein attributed his discoveries, and which popular accounts took as one of the givens of an Einsteinian universe, was associated with "the American love of free expression," that is, with political ideals and institutions.[28] Writing about Yeats, Edmund Wilson noted that science and democracy were associated in part simply because their rise coincided chronologically; Yeats, he wrote, stood "apart from the democratic, the scientific, modern world" (p. 40). Drawing on a slightly more detailed picture of the new physics, Wyndham Lewis went further, asserting that Einsteinian physics fostered a picture of "an equal reality in everything, a democratically distributed reality, as it were."[29]

As Moore's poem reveals, American modernists were quick to seize on hints that modern science might be recruited in the service of America and further might serve, however illegitimately, to connect the nation's literary and political experiments. Williams's note for a college lecture given in 1957 seems to rest on such a view. He wrote that "influenced perhaps by Albert Einstein's work in the field of theoretical physics, the American Idiom has grown to be a unique modern language."[30] For Williams, relativity theory seemed to sanction American speech because both had been called democratic and modern. Moreover, because of this association, modern science seemed to bless Williams's search for "a new measure or a new way of measuring that [would] be commensurate with the social, economic world in which we are living" (*SE* 283).

In short, Einstein's theories helped poetry to claim it had a function specifically in the real world of American democracy and so enabled American poets to feel rooted in their society. Ironically, of course, the new physics might be said to have called into question the solidity of the everyday world, and to have violated common sense notions of what reality is.[31] It is precisely this desolidification that Lewis calls democratic. Still, by a poetic logic, Williams seems to have taken Einstein's *democratized universe*—"an organic structure, in which all are interdependent"—as legitimizing his ordinary American subject matter and poetry's place in America.[32]

Relativity theory was also, needless to say, connected with time and motion, a connection to which Williams refers in jotting notes on "pure time—a motion. (chemistry and physics: one.)" (*EK* 111). This emphasis on speed and motion was also quickly associated with America, in part because of the speed of American trains and motor cars, but also because of modes of production identified with Ford and with Taylor's time studies or *scientific management*.[33] Stuart Chase's 1929 series of articles in the *New Republic*, for instance, cited Dewey's claim that American "civilization is like a Ford . . . restless, aimless, but vital and moving," and went on to discuss labor and production practices.[34] By 1928, *transition's* American issue contained an article by Max Rychner cautioning that America might remember that its real strength was technology, while the scientists it wanted to claim—Gauss and Einstein, for example—were Europeans.[35]

By the 1920s, nonetheless, Einstein was not only adopted as a modern American, he was also, at least indirectly, linked with the mistrust of American industrial modernity. In the January 1921 issue of his magazine, *Contact*, for example, Williams declared St. Francis "patron saint of the United States" (*SE* 27). In the summer of 1921, also in *Contact*, he associated Einstein with St. Francis in a poem called "St. Francis Einstein of the Daffodils." The connection is less peculiar than it first appears. As Jackson Lears points out in *No Place of Grace*, suspicions of urban industrial culture, even on the part of those who admired American technology per se, engendered a revolt, "often in the name of a vitalist cult of energy and process; and a parallel recovery of the primal, irrational forces in the human psyche," popularly associated with a romanticized medievalism.[36] Bergson, an outspoken irrationalist and foe of positivism, helped link Einstein in the popular imagination with the cult of vitalism and so, in Williams's mind, with St. Francis.

In a 1946 poem on Williams, Kenneth Rexroth mused: "Sometimes I think you are like / St. Francis,"[37] drawing not only on Williams's poem about St. Francis Einstein, but also on the common image of St. Francis's sympathetic understanding of all elements of his world. In *Exile's Return,* Malcolm Cowley recalled that his generation felt alienated from modern society and that authors such as Gissing had identified science as one of the causes of modern alienation.[38] Ironically, then, those who mistrusted science often cited Einstein, who was seen as having reinstated a more human-centered universe. Thus, Wyndham Lewis's somewhat cranky description of relativity theory as "the physics of the primitive mind, . . . of the *naïf*" refers to the association of the irrational and the pre-modern with Einsteinian physics.[39]

The idea that modern science might overcome social alienation was tied, as well, to the belief that relativity theory rescued humanity from a deterministic world and thereby restored the central role of free will and the human imagination, offering a universe in which poets could claim that their work mattered. Whitehead's writings encouraged these beliefs, proposing that the ancient world took its stand on the drama of the universe, "the modern world upon the inward drama of the Soul."[40] The idea that the mechanistic universe had been replaced with a human-centered universe became part of popular mythology. As Williams put it, the measurements of aesthetics, science, morality, or religion became "one in that they are the functions of one thing: man."[41] Edmund Wilson repeats the notion in *Axel's Castle:* "The scientific point of view no longer implies . . . determinism. . . . Man is no longer to be a tiny exile . . . what he has been thinking of as his soul, the exclusive possession of human beings, is somehow bound up with that external nature which he has been regarding as inanimate or alien, and his mind . . . turns out to have constructed its own universe" (pp. 90–91). Given Eddington's often reprinted comment on the footprint discovered by science being *our own,* or Heisenberg's notion that we constitute what we observe, or Steinmetz's remarks on how "our mind clothes the events of nature," the new physics seemed to heal the gap between the subjective and objective worlds and thus to validate once seemingly less objective disciplines such as art and poetry.[42]

As these ideas about physics and free will reveal, Einstein's theories were often mixed by writers such as Williams and Wilson with other scientific discoveries. In part because of his own impressive achievement and in part through the popular association of a universe of flux with Bergsonian notions of creative evolution, Einstein was thought to "have revealed to the imagination a new flexibility and freedom," as Wilson wrote in *Axel's Castle* (p. 298). It was, of course, quantum mechanics and Heisenberg's principle of indeterminancy, which Einstein found "unnatural" and even Planck disliked, that actually raised questions about causality and, by extension, about determinism and free will.[43] Heisenberg himself noted that his discoveries cast doubt on the objectivity of classical physics: "It may be said that classical physics is just that idealization in which we can speak about parts of the world without any reference to ourselves."[44] Although popular sources were more enthusiastic in declaring that free will had been restored, the physicists also helped to spread the idea that here too physics had something to say to poets and philosophers. In 1934, Bohr wrote: "Just as the freedom of the will is an experiential category of our psychic life, causality may be considered as a mode of perception by which we reduce our sense impressions to order."[45]

The physicists may well have been wrong in thinking that their findings settled questions about determinism and free will, questions that were raised for them by scientific problems.[46] In his 1941 *Let The People Think*, Bertrand Russell pointed out that scientific problems led scientists to take an interest in philosophy: "This is especially true of the theory of relativity, with its merging of space and time into the single space-time order of events. But it is true also of the theory of quanta, with its apparent need of discontinuous motion."[47] Implicit in Russell's remarks is the recognition that the new physics was not a monolith and that different discoveries engendered opposing and sometimes incompatible philosophies, thereby calling into question any single view that drew exclusively on Einstein's vision of a continuum in the universe or on quantum mechanical discontinuities. Moreover, scientists did not necessarily think their theories were related to poetry. Heisenberg explained that any "kind of understanding, scientific or not, depends on our language, on the communication of ideas. Every description of phenomena, of experiments and their results, rests upon language as the only means of communication."[48] But he was also capable of faulting other scientists such as Bohr for using "ambiguous . . . language," which "reminds us of a similar use of the language in daily life or in poetry."[49]

Writers, nonetheless, were drawing on speculations that derived from quantum mechanics and relativity theory as well as the Curies' work, which most laymen did not distinguish as separate, as when Williams described poetry as a *force* or insisted that poetry, sanctioned by science, "returns authority to man."[50] Even as late as 1960, Williams was still mixing sources, as in his explanation of his attempt to "bring in the idea of mathematics. Of Einstein. Not Einstein, we'll say, but Einstein's ideas. The uncertainty of space" (*SSA* 45).

As Williams's own uncertainty makes clear, most poets did not read the scientists themselves. Marianne Moore read reviews of Einstein's biography, as well as the *Scientific Monthly* and several books by Russell.[51] Russell, Whitehead, and Samuel Alexander were probably the most widely read popularizers; Stevens's notebooks and essays indicate that he read, or at least read about, Alexander, who was cited also by Lewis in *Time and Western Man*, and whose Gifford lectures for 1916–1918 formed the basis of his 1920 book, *Space, Time, and Deity*.[52] Drawing on a combination of readings from Bergson, Minkowski, and others, Alexander proposed that motion or *pure events* made of space-time were the basic stuff of the universe from which all else, including consciousness and—ultimately—deity, evolves.[53]

Just as, it seems, Williams's information about the Curies came from the 1938 translation of Eve Curie's biography of her mother and from MGM's 1944 *Madame Curie*, so too many American poets who used the vocabulary of the new science may not have known even the popularizers' books at first.[54] Williams's poem on Einstein, for example, appeared in *Contact* in 1921. In it are images of force, light, and relative measurement that, as Carol Donley has shown, would have been familiar to any reader of the *New York Times*.[55] John Riordan introduced Williams to more accurate accounts of the relativity theory five years later.[56] And it was not until 1927 that Williams announced he had read Whitehead (*SL* 79). In other words, Williams's excitement about Einstein came well before he had read about Einstein's work in any detail. Similarly, although Stevens's library included *Axel's Castle,* as well as a few other books that mentioned modern science, he seems first to have become aware of the specific implications of the new physics only after reading C. E. M. Joad's 1940 and 1941 articles on Bergson and Alexander in the *New Statesman.*[57]

As R. P. Blackmur pointed out in his 1928 review of *Time and Western Man* for *Hound & Horn,* by the late 1920s Lewis was already warning against the "uncritical analogues of science" or the "conversion [of physics] to other fields."[58] The language used by the scientists themselves probably encouraged this tendency; at least to the layman it seemed, as Hart Crane noted in 1926, that scientists were "proceeding to measure the universe on principles . . . quite as metaphorical . . . as some of the axioms in Job."[59] In any case, Einsteinian physics certainly gathered more popularity than understanding.

There were cautions issued in some of the more popular accounts of the new science. In *Science and the Modern World,* Whitehead noted and warned against the "tendency to give an extreme subjectivist interpretation to this new doctrine."[60] In particular, such interpretations came from readers of Bergson "whose ideas [were] . . . restored to great honour . . . by relativity in its literary or non-scientific capacity."[61] Russell described Bergson's account of memory as "a poetic way of speaking, but . . . [not] a scientifically accurate way of stating the facts."[62] Nonetheless, as if describing matters of scientific fact, many popular books and articles stressed poetic and literary analogues of the new physics.

Morris Cohen's article in the *New Republic,* a review of no less than eleven recent books on the new physics, deplored especially one book which announced that "the theory of relativity demands a Bergsonian philosophy."[63] Cohen's article contained one of the more detailed and serious discussions of Einsteinian physics available to those who had not read the books being reviewed. It also neatly outlined the current misconceptions about relativity theory and made clear how poets might have seen Einsteinian physics, and Einstein's achievement, as proof of the value and validity of the poetry they were already writing. There was a place ready made for Einstein's theory, especially in America where artists and writers were casting about for a way to elevate their more human enterprises and where prior associations between native American energy, organicism, democratic ideals, and a national taste for speed made popular accounts of the new physics appear to offer a natural and authoritative defense of the arts.[64]

At times, as Williams for example acknowledges, the imagist aesthetic so often associated with technology was superseded by the new model provided by science. In 1948, in "The Poem as a Field of Action"—the title drawing again on the vocabulary of physics—Williams announced that relativity theory dictated a new structure for poetry, according to which "there is no such thing as 'free verse'. . . . Imagism was not structural: that was the reason for its disappearance" (*SE* 283). Well before he developed his triadic *variable foot*, Williams insisted that Einstein's theories demanded a new style. Despite Williams's insistence on structure, it was motion or flux to which the analogy with physics drew attention.

As Lewis pointed out, the new physics was associated with inner as well as outer flux. The association, though implicit, is suggested in *Science and the Modern World* by Whitehead's mention of mental change and the vocabulary of physics in his discussion of the transience of the field of perception and the psychological field. Noting that "the mind is the major permanence, permeating that complete field, whose endurance is the living soul," Whitehead concludes that this "soul cries aloud for release into change. . . . The transitions of humor, wit, irreverence, play, sleep, and—above all—of art are necessary for it."[65] In short, as with Einstein's constant, the mind in motion was claimed as a universal truth. Ignoring the actual differences between the poetics and the practice of writers such as Stein, Joyce, and (more implicitly) Whitman, commentators tended to lump any style that seemed to stress mental motion and playfulness with Einsteinian physics and with a stream of consciousness.[66]

Although he himself would later link Einstein's theories with a more structured style, many of Williams's early and middle poetic experiments call attention to imaginative linguistic play and mental leaps of association, a style Lewis described as inspired by the "flux of Bergson, with its Time-god, and the einsteinian flux, with its god, Space-time."[67] In "The Trees," for example, Williams writes:

> Christ, the bastards
> haven't even sense enough
> to stay out in the rain—
>
> Wha ha ha ha
>
> Wheeeeee
>
> [*CEP* 66]

As Henry Sayre has argued, this is Williams's loose or visionary style. The image of the trees "is graced by . . . subjective presence";[68] the stream of associations, inversions of idiom ("sense enough / to stay out in the rain") and nonsense syllables suggest the imagination's rapid motion and playfulness. Such inner motion and multiplicity was identified with the reality described by the new physics as reported in books like Alexander's *Space, Time, and Deity* or Wilson's *Axel's Castle*.

A similar appeal to the sounds and playfulness of interior monologues is found in Joyce's *Ulysses*, as in this passage from the "Proteus" section of the novel:

> I am, a stride at a time. A very short space of time through very short times of space. Five, six: the *nacheinander*. Exactly: and that is the ineluctable modality of the audible. Open your eyes. No. Jesus! If I fell over a cliff that beetles o'er his base, fell through the *nebeneinander* ineluctably. I am getting on nicely in the dark. My ash sword hangs at my side. Tap with it: they do. My two feet in his boots are at the end of his legs, *nebeneinander*. Sounds solid: made by the mallet of *Los Demiurgos*. Am I walking into eternity along Sandymount strand? Crush, crack, crick, crick.[69]

Referring to space, time, and causality, such passages provided evidence for those like Wilson or Lewis who linked Joyce's writings with the new physics. More importantly, the style of the passage calls attention to the mind's protean flux, wherein sheer sound ("crack, crick"), literary associations, and puns ("sounds solid") follow one another in rapid succession. Williams explicitly associated this Joycean style with the new physics: "In . . . the form of the line (of which diction is a part) it must have room for the best of Joyce—the best of all living . . . thought and—Einstein" (*SL* 135). Wyndham

Lewis more generally linked the *gospel of action* with Stein, Joyce, Lawrence, Bergson, Einstein, and most Americans.[70]

Those poets who experimented with styles that could be described as emphasizing thought in motion—whether Stevens's meditative line or Williams's more surrealistic experimental books—at one time or another associated their style with Whitman. In his *Autobiography*, Williams described much of his early journal writing as "Whitmanesque" (A 53), and in at least one poem Stevens explicitly attributes the process so often discussed and exemplified in his poetry to Whitman (*CP* 150).[71]

By the 1920s, for Williams and somewhat later for Stevens, the analogy with physics was used to defend their Whitmanian style and seems to have further strengthened their commitment to such a style, which came to be seen, as Williams's or Lewis's statements indicate, as authorized in part by modern science. Drawing on his own recent experiments with the American idiom and with a stream-of-consciousness narrative, Williams described his vision of modern poetry to Kay Boyle in 1932: "The line must be pliable with speech, for speech, for thought, for the intricacies of new thought--the universality of science compels it" (*SL* 134–35).

Similarly, Stevens's earliest, anecdotal poems were viewed by the critics to whom he responded most (though most defensively) as overly precious gemlike objects. In his 1935 review of *Ideas of Order* and *Harmonium*, for example, Stanley Burnshaw called Stevens's work " 'scientific,' objectified . . . crystallography."[72] By 1940, based on his understanding of modern physics, Stevens was writing and defending the more meditative style of his middle and late poetry. Moore's models seem more Emersonian and appeal more to biological science, but the mix produced similar results in terms of a poetics centered upon the mind's encounters with the world, on singing one's self, rather than on the poem as an object comparable to or about other physical objects.[73]

The vision of poems as emblems and records of, as well as occasions for, mental energy does not entirely accord with the vision of poems that I have already articulated. The two models of poetry were not exclusively associated with, and certainly not caused by, the use of analogies with technological products or with the new physics. Henry

Sayre describes how Williams, for example, from very early in his career and well before he began alluding to the new physics, wrote two different kinds of poetry, one visual and one visionary, betraying a "fundamental aesthetic inconsistency."[74] Moreover, the tension between two different aesthetics and their accompanying poetics was rarely obvious to the poets. Rather, the use of analogies from technology and physics—given the usual conflation of the two and given that both were at times linked with creative energy, at times with precision—probably helped obscure the discrepancy between a machine aesthetic and an action aesthetic for most writers.

Yet when Williams praises a poem as "a thing" (writing to Zukofsky in 1928, *SL* 94), he seems to be appealing to a vision of a poem being, like a machine, a carefully structured, manufactured object, resulting from the poet's close act of attention to the world (and calling forth from the reader a parallel act of attention). The model is, among other things, visual; as Williams wrote, again to Zukofsky: "Eyes have always stood first in the poet's equipment" (*SL* 102).[75] In roughly the same period, however, Williams also celebrated himself as being drunk on "words at flow" (*SL* 175), recalling his description of science as being "drunk with work . . . the exhalation of a fever" (*VP* 114). Poetry is thus envisioned as a record of an internal and often disorderly reality, or the "flight of emblemata through [the] . . . mind" (*OP* 71), as Stevens put it. The poem becomes a process of mental and verbal exploration. The model is further associated with physics in Williams's statement about how poetry cannot be written according to a plan: "Instincts are the feelers into new territory—even Einstein has recently acknowledged or stated that" (*SL* 252). Lewis emphasizes the distinction, calling the Einsteinian poetics of action, with its interest in a psychological flux, too abstract and arguing for his own commitment to a more visual aesthetic: "I am for the physical world."[76]

A poem Williams first published in 1940 again helps clarify what was meant by a poetry of process. Williams focuses on the mind moving "after a pattern / which is the mind itself, turning / and twisting the theme" (*CLP* 12). The reality to which the poem refers, and about which it talks, is internal or psychological. If Williams insists here on structure or patterns, he still conceptualizes such compositions as being "without code," as the free play of the mind. Indeed, the poem insists: "We are not here, you understand, / but in the mind, that circumstance / of which the speech is poetry" (*CLP* 12). The style of the poem, too, stresses a poetics of process, incorporating Whitmanian lists—"In your minds you jump from doors / to sad departings, pigeons, dreams / of terror, to cathedrals" (*CLP* 12)—contained within the mind of a speaker whose movement of thought again suggests more free association than visualization.

Even the title of the poem, "Writer's Prologue to a Play in Verse," stresses the characteristics most often defended by appeal to a physics that proclaimed "*universals . . . have . . . the form of motion*."[77] The poem is a prologue, emphasizing its lack of finality and the exploratory nature of the monologue; the reference to a "play" is both an insistence on dramatic action and on the free play of the mind. Finally, of course, the pun on play is itself part of the play.

That Williams included this poem in *The Wedge* (1944), the introduction to which defines a poem as a "machine made of words" (*CLP* 4), is characteristic; the two poetics discussed here were not presented or thought of as distinct. Indeed, Williams's draft for an article on objectivism for the *Princeton Encyclopedia of Poetry and Poetics* referred to both models. The article described a poem as "an object to be dealt with as such. . . . with a special eye to its structural aspect, how it ha[s] been constructed," but Williams added: it "arose as an aftermath of Imagism. . . . The mind rather than the unsupported eye entered . . . the picture." Williams's dissatisfaction with this attempt to reconcile his two aesthetics is clear in his indication that the "movement, never widely accepted, was early abandoned."[78]

Ultimately, however, many poets seem to have felt that the analogy with physics allowed them to reconcile their two styles—the style of their imagistic, object-like poems and that of their more psychological, Whitmanian poems—by appeal to a universe in which objects were part of a universal, inner and outer, flux. Even Crane's bridge may be depicted as a segment of space-time, allowing him to reconcile his distaste for urban technology with his desire for a modern myth; the poem seems to draw on the Einsteinian picture of a cosmos of finite but unbounded space: "that star-glistered salver of infinity, / The circle, blind crucible of endless space, / Is sluiced by motion,—

subjugated never."[79] In the new picture revealed by science, objects were neither static nor atomistic. Instead, as Perry Hobbs's 1927 article for *Hound & Horn* explained, objects were, like all matter, "made up . . . of steady events, rhythms, and transactions."[80] The poetic view of a central, dynamic relationship between observer and observed to which the poets were attracted in part because of their legacy from Whitman and Emerson, was sanctioned by science as the best available picture of objective reality.

Despite potential internal contradiction, then, most poets continued either unthinkingly to mix their two aesthetics or to assume that Einsteinian physics reconciled the two models of poetry by addressing the question of the relationship between objects and mental motions. Moreover, since physics seemed to mandate a style already associated with Whitman and so with America, the appeal of such a style was double, satisfying the poets' desire to prove that America could produce strong modern poetry and seeming to ground modernist poetry in the *reality* connected, through science and technology, with America. Not only did claiming the reality found in poetry (or enacted in the poem) was sanctioned by physics open a potential place for poetry in modern America, but the popular respect for Einstein's theoretical science in America seemed to guarantee a respect for pure creativity, rather than for the production of some commodity. Finally, if imagination was at issue, the poets could claim that they were the true authorities and most deserving of public approval. As Williams said, "The imagination is recognizable at the peak of all professions but in the arts it exists at its purest."[81]

The use of analogies with physics, in short, often allowed poets to overcome or overlook their commitments to potentially opposed aesthetics. As a way of envisioning a place for poetry in modern American society, however, the analogy raised other problems, as can be seen by looking at a 1926 poem by Robert McAlmon. *The Portrait of a Generation*, a copy of which Williams presented to Stevens, describes a "specialist in relativity / who spoke of Einstein's dynamic eyes."[82] The appeal to dynamism is again to the universe of flux and motion and also to the relative nature of observation Einstein was understood to have verified; the classic example of an observer on a moving train (which would have been familiar from any number of popular accounts) probably informs the description of Einstein's eyes as dynamic.[83] Here too is an appeal to Einstein's own creativity, another conflation of creator and created that encouraged the popular misunderstanding of relativity as descriptive of a Bergsonian inner flux. Finally, McAlmon's portrait concludes: "He spoke of music / which so well becomes a man / Of mathematical talents / He might give up science to compose" (p. 39).

Einstein may have disrupted certain classical notions about the true nature of energy, but the image suggests not only that any given individual has limited energy, needing to pick one area of expertise, but also that, within modern society, the arts and sciences might still be at odds. Literally, and discounting for the moment the social satire involved, McAlmon suggests that, like Einstein with his violin, others trained in science might easily take up art. The suggestion turns on its ear a protest made by any number of writers, namely that artists all along intuitively knew what Einstein had discovered, and they knew it first. A 1919 editorial in the *Nation* reads: "Religion and poetry, of course, have always known that space and time are relative."[84] Whitehead's chapter on the Romantic poets proceeds along similar lines, arguing that the poets' early sense of an organic unity in the world was only later arrived at by science. Inadvertently, however, McAlmon notes that if one grants science and the arts shared insights, the result should be that scientists can become artists as easily as artists can lay claim to having had early insights into scientific facts.

As McAlmon's description of the scientist turned composer shows, poets were as defensive about science as about technology. Other examples are easily found. In 1933, even as he was citing Einstein, Williams wrote: "From knowledge possessed by a man springs poetry. From science springs the machine" (*SL* 137).[85] And Crane protested that the "truth which science pursues is radically different from the . . . 'truth' of the poet."[86] The desire to preserve the seriousness and unique importance of poetry even while using modern science and mathematics to stress poetry's validity is starkly presented in two late remarks by Williams. In 1955 Williams wrote: "The variability of all the values must be recognized in our verse, mathematics itself demands it of us."[87] Yet in roughly the same period, speaking to an audience at Washington University, he explained that mathematicians "meant nothing to me, I acknowledged them, of course, but refused to acknowledge, that in all their ranks was an individual that surpassed the artist."[88] In short, adopting science or scientists was a double-edged strategy, at times seeming more likely to threaten poetry than to strengthen its image.

Despite such muted ambivalence, invocations of Einstein in general were designed to provide an authoritative foundation for poetry. Allen Tate's acerbic comment on *Axel's Castle* missed the point: "Why suppose," he asked, "that science, rather than a better income, or a trip to Europe, is what poetry needs?"[89] Given the universal appeals to science, especially in America, however, the question might better be: Why did poetry think it needed science? The answer is in part that science, like American technology, offered American poets a way of defending their poetry in a culture that had no clearly defined place for poetry, but did respect the new physics. The poets who in many ways were very much part of their culture usually shared that respect, and shared as well the respect for what was grounded in the real world.

As Williams wrote in 1956, strongly emphasizing his need for a foundation for his poetics: "The variable foot is only valid as it is part of the new world—of Einstein and the others."[90] However vague or misguided the uses of the new physics were, however many contradictions or problems were raised, the appeal to Einsteinian physics allowed poets to claim, with Edmund Wilson, that they were part of "a revolution analogous to that which ha[d] taken place in science" or even, drawing on suggestions like Bertrand Russell's that Einsteinian physics "demanded . . . a change in our imaginative picture of the world," to claim that science needed poets.[91] In short, defenses of poetry supported by analogies with physics were critical for the development of American modernist poetry and poetics, and especially for two of the poets whose work is at the core of this book.

Moore, Williams, and Stevens differ in many ways. Stevens, for instance, is most often characterized as a poet of "the act of the mind" (*CP* 240), interested more in a poetics of process and less in the objective world than in the mind's confrontations with reality. Moore and Williams, on the other hand, are most often called *objectivists*, although Moore seems most interested in "the relations between aesthetic integrity and moral integrity," while, especially in the better known examples of his early work, Williams's attention is to objects themselves rather than to the moral qualities they figure.[92] Nonetheless, all three are similar in their acute awareness of the position of poetry in America and in their insistence that poetry address the collective needs of Americans. Thus, in their attempts to define modern poetry's status, all three drew on analogies with science or technology in order to claim that poetry was, like these other disciplines, important to America. The chameleon-like strategy of saying poetry resembles other more widely respected fields may not ultimately be convincing; however, as Williams, Moore, and Stevens each confronted the ambivalences and tensions generally involved in such defenses of poetry, their sense of poetry's importance and their own writing was strengthened.

4

William Carlos Williams: There's More Than Just One Kind of Grace

WILLIAM Carlos Williams's connections to other modern artists and writers have been well documented, and recent criticism has begun to notice his use of Einstein and the Curies in his later poetry.[1] There is also a growing interest in the role that popular debates and misconceptions about science and technology played in Williams's evolving definition of literary modernism.[2] Throughout his career, Williams's interests in these other fields were related to the development of his views on the place and function of poetry in America. As he asks in his 1914 "The Wanderer," the poem with which he chose to open his *Collected Earlier Poems*, "How shall I be a mirror to this modernity?" (*CEP* 3).

In his *Autobiography*, Williams admits to having been unclear at first about what, exactly, influenced his early modernist experiments: "What were we seeking? No one knew consistently enough to formulate a 'movement.' We were . . . closely allied with the painters. . . . We had followed Pound's instructions" (A 148). As Williams confesses here, his earliest poetry appears sometimes to rest its case for being "a mirror to . . . modernity" simply on its association with other, well-publicized, avant-garde movements in the arts. Yet Williams's early theoretical writing also reveals that he was asking sharp-sighted questions about the nature of modernity in American society and its relationship to modernism.

For many in the first decades of this century modernity and progress were linked to advances in technology, popularity conflated with science. Leo Stein's 1917 article, "American Optimism," explains, for example, that "machinery and innumerable substances of value to industry became rapidly available, and . . . the advance in the mastery and control of material conditions became orderly and continuous."[3] Stein suggests that the confidence bred by technological discoveries is false, in that it ignores the problem of which values will inform the direction taken by technology, and he proposes science as a source of the reforms he finds necessary. The article finally calls for a social science to be founded by men comparable to Lister or Pasteur and to revitalize science and the arts alike by providing a "hygiene of the soul."[4] Although Stein begins by distinguishing between the theoretical achievements of the scientist and the practical achievements of the engineer, his final analogy between the practical hygienic advancements provided by, as he says, science and the spiritual hygiene needed by society blurs the distinction between science and technology.

It is illuminating to juxtapose Stein's optimistic call for a reform of American values with another view published the same month. Just as Stein's article appeared, Duchamp resigned from the organizing committee for the Independents Exhibition in New York. Despite their announced policy of deploring censorship, the committee refused to accept the porcelain urinal that Duchamp anonymously submitted to the show. In the May 1917 issue of *The Blind Man*, Duchamp pronounced judgment on American taste: "The only works of art America has given are her plumbing and her bridges."[5] Duchamp's admiration for the look of American plumbing and bridges was genuine; his remarks, however, were intended to shock his audience. Those who refused to accept his urinal could hardly have been expected to agree with his judgment on American plumbing; celebrations of American technology and hygiene did not usually view technological products as artistic per se. As Duchamp's dispute with the Independents suggests, it was not clear how art might participate in the modernity for which hygiene stood.

Stein's and Duchamp's articles reveal two radically different ways in which American technology was viewed. On the one hand, technological progress and hygiene seemed symptomatic of a system of values, which included materialism, efficiency, and perhaps squeamishness (often attributed to Puritanism), that was unlikely to provide an environment in which the arts might flourish.[6] Technology, on the other hand, seemed to be the distinguishing positive feature of modernity that the arts wished to claim.

Any reader of *The Blind Man* and the *Seven Arts* in May of 1917 might have been struck by these ironically juxtaposed concepts of American hygiene. Williams read both journals and echoed Duchamp, whom he admired, in his later references to "American plumbing [and] . . . American bridges" (*SE* 35).[7] It is pleasing to imagine that the shock of reading Duchamp and Stein together informs Williams's later ambivalent references to the hygienic inventions of modern America, its "bathrooms, kitchens, [and] hospitals" (*IAG* 177) or "modern plumbing [and] refrigeration" (*SE* 176). Williams's use of such images, however, probably stems from more direct, personal experience. As a doctor, he saw the value of hygiene and well-equipped American hospitals. The material progress of America was, Williams knew, genuine cause for optimism. Yet, as Henry Sayre has documented, by 1921, when most of the writers and artists whom he knew had left for Europe, Williams also recognized the partial truth of Marsden Hartley's perception that America had few good artists because "the integrity of the artist [was] trifled with by the intriguing agencies of materialism."[8]

There are, of course, differences between material progress, materialism, and, for that matter, materiality as understood by modern painters such as Hartley, Alanson Hartpence, and Charles Demuth.[9] Nonetheless, as Hartley's comments about artistic integrity indicate, the American context often blurred the distinction between art objects that insisted on the materiality of art, material inventions, and a materialistic interest in commercial products. In thus appealing to one aspect of American technology (such as a machine aesthetic or even material progress based on technological breakthroughs), artists and writers risked having their work judged by the same criteria used to evaluate consumer products or technological inventions.

Williams was aware of judgments like Hartley's when—from the early volumes of poetry, *Sour Grapes* and *Spring and All*, through the 1934 volume of *Collected Poems*, published by the Objectivist Press, and the 1935 *An Early Martyr*—he drew on his alliance with modern artists and architects in his experiments with a machine aesthetic.[10] Indeed, he is probably best known for the sparse, hard-edged use of language found in early poems such as "The Red Wheelbarrow," "The Locust Tree in Flower," or "Young Sycamore."

Williams's acute awareness of the American public that Hartley described is revealed in remarks like one in *The Embodiment of Knowledge*, from the late 1920s, where he mentions the need to remove the film from people's eyes. This remark is followed by discussions of silver nitrate, which helped save infants' eyesight and so was a good example of material progress; of poetry's ability to give people contact with their world, which involved the kind of materiality and seeing anew that Demuth and others celebrated; and of commodities sold to improve personal hygiene, which involved American materialism and consumption. Aptly, when he moves from discussing the literal and poetic changes that might help Americans see to discussing American consumers, Williams shifts from images of eyes to an image of teeth and toothpaste. In responding to characteristic American attitudes ("The U.S., land of the film. Use a good toothpaste"), he sees he must insist that "poetry is something else" (*EK* 27), that is, that poetry is not like toothpaste. Williams's humor does not mask his recognition that American hygiene was usually associated with consumer products and materialism.

Despite his critiques of America, Williams did not follow Pound and others to Europe. He thus needed to defend his allegiance to a culture that, from one point of view, could not understand, let alone foster, experiments in the arts. Specifically, he felt the need to argue that his life as a doctor in America and his daily contact with people who were neither literary nor avant-garde formed the best possible environment in which to develop a new poetry.[11] At the same time, many of the aspects of popular American culture that Williams defended, especially its attention to practical matters and its sometimes naive optimism, seemed to have caused the alienation of the expatriates to whom Williams addressed his defense.

A sometimes qualified optimism mixed with a concern for the social role of poetry occurs frequently in American modernist poets: There is Moore's acknowledgement, in "Poetry," that there is something important and useful in poetry (*Comp* 266–67); Stevens's insistence that poetry helps "people to live their lives" (*NA* 29); and Williams's late claim that, although it "is difficult / to get the news from poems / . . . men die miserably every day / for lack / of what is found there" (*PB* 161–62). Hart Crane often voiced his belief that poetry had a universal and purely spiritual function; yet even Crane, whose poems are usually seen as not deeply rooted in either history or the contemporary world, said that he was, as a poet, "concerned with the future of America . . . because . . . here [were] destined to be discovered certain as yet undefined spiritual quantities. . . . And in this process [he felt himself to be] a potential factor."[12]

The idea that Americans were optimists was commonplace. Man Ray suggested, for example, that the nihilism of European Dadaism could not be transported to America: "You can't adapt it to America, it doesn't work."[13] And as early as 1908, in *The Wine of the Puritans*, Van Wyck Brooks wrote that Americans did not have "the pessimism of Europe."[14] The causes of this optimism are clearly more complex than Stein, who related it to technological progress, suggested. Stein's suggestion, made in 1917, does nevertheless reveal a contemporary view that American optimism stemmed from faith in material progress.[15]

Certainly Williams's defense of America as the birthplace of modernism rests on his awareness that America was distinctively modern because of its technology and its industrial know-how. For example, in 1921 he defended practical American ingenuity as that which could produce a distinctive American poetry:

> Not that Americans today can be anything less than citizens of the world; but being inclined to run off to London and Paris it is inexplicable that in every case they have forgotten or not known that the experience of native local contacts, which they take with them, is the only thing that can give that differentiated quality of presentation to their work which at first enriches their new sphere and later alone might carry them far as creative artists in the continental hurly-burly. Pound ran to Europe in a hurry. It is

> understandable. But he had not sufficient ground to stand on for more than perhaps two years. . . . It has been by paying naked attention first to the thing itself that American plumbing . . . indexing systems, locomotives, printing presses, city buildings, farm implements and a thousand other things have become notable in the world. Yet we are timid in believing that in the arts discovery and invention will take the same course. And there is no reason why they should unless our writers have the inventive intelligence of our engineers and cobblers. [*SE* 35][16]

Williams thus defends American modernism by appealing to American productivity in other fields, including engineering.[17] But he also came to see the problem this posed. When Stein called for a social science or Frank A. Manny, whose article appeared in the June 1917 issue of the *Seven Arts*, concluded that art "like industry must achieve visible and tangible results which minister to human use," their language proposed that the hard results of technology be taken as the measure of advances in the arts.[18] As an American and a doctor, Williams too could be enthusiastic about technological advances, but he knew also that poetry could not easily be defended by appealing to its tangible results.

Thus, when Williams allied himself with the practical American vision represented by Stein and Manny, he generally did so ironically. In his *Autobiography*, looking back on his trip abroad in the mid-1920s, he has a young French salesman admit that French sales "methods are not as quick or efficient as yours" (A 192). In the same chapter of the *Autobiography*, Williams presents himself as a practical, unsophisticated American asked to make a speech at a dinner party in Paris that included such giants of European modernism as Joyce, Antheil, and Duchamp, as well as the man Williams would later count as a close friend, Ford Madox Ford. Williams writes: "What had I to say with all eyes, especially those of the Frenchmen, gimleted upon me to see what this American could possibly signify, if anything? I had nothing in common with them. I . . . sat down feeling like a fool" (A 195).[19]

While Williams's pose as an outsider, a naif, is ironic, the irony is uneasy. If practical America held the key to the modern world, Ford and others might be right that practical America, which measured success by utility or financial gain, was no place for poets. The French, especially, might admire the look of American modernity, but they did not expect America to produce modern writers.

Nor was there an audience for modernist work in the United States. Williams registered this fact in his poems as well as his prose. His 1916 poem, "Pastoral" ("When I was younger"), displays the hard-edged, clean, and accurate modern style associated with American technology.[20] The poem ends after having described the beauty of a poor neighborhood: "No one / will believe this / of vast import to the nation" (*CEP* 121). Literally, what Williams finds important is the look—the lines and colors—of the urban landscape, although implicitly the look of his poem on the page is also part of the machine aesthetic he praises.[21] As Marjorie Perloff, among others, has argued, Williams's visual form was influenced by paintings including Picabia's 1915 object-portraits, which in turn drew on newspaper advertisements and mail order catalogue drawings of merchandise, that is, on uniquely American lines associated with American products.[22] Although Williams, like the artists he admired, adopted what was seen as a specifically American aesthetic, the America he claimed as his own was primarily interested in quite another kind of import, with which poetry, practically speaking, could not compete. "Pastoral" challenges American values with its pun on the word import, and exemplifies one strategy Williams used to redefine the significance of American technology. Yet his skepticism about whether or not the American public would accept his challenge is clear.

Williams also proposed another view of what technology reveals about the American character, one which tried to avoid comparisons between the tangible or profitable results of technology and the results of poetry. In suggesting that poets "take the same course" as engineers, Williams not only compiles a long list of successful inventions by American engineers, but he also celebrates the inventive intelligence or imaginative energy that produced such goods. Both Henry Sayre and Cecelia Tichi have recently noted how often Williams describes his own enterprise as like that of an engineer in that he emphasizes the inventive assembling of prefabricated parts.[23] Indeed, Williams was not alone in finding a creative spirit behind the often dismal results of industrial growth. The *Seven Arts*, which printed both Stein's and

Manny's articles, also presented Bertrand Russell's declaration that one can see "fundamental [that is, creative] work" in science, as well as James Oppenheim's description of the "glow of the human spirit in the laboratory."[24]

This reevaluation of technology and science is important: A technological society may appear at first glance to have no use for poetry, but on closer inspection, technological science is seen to require creativity. By not being tied to commerce, and precisely by not producing results like factories and plastics, poetry offers what is essentially valuable in modern technology, but in purer form, as it were. Williams thus stresses the pure creativity of poetry: "imagination . . . exists . . . in poetry at its most articulate and impassioned."[25]

Williams's emphasis on inventiveness was not at odds with his attraction to a machine aesthetic, but it did entail making careful distinctions between the products of American technology and the spirit that produced them, as is shown in "The Basis of Faith in Art." In this 1937 dialogue with his brother concerning the relationship between architecture and poetry, Williams observes that "in a scientific era . . . the artist's protest that his art is wholly nonutilitarian has a certain amount of truth in it. . . . The uselessness of it might constitute its principal use, sometimes" (*SE* 179). His point is that art is important, but its usefulness cannot be taken as analogous to the uses of electricity, "modern plumbing, refrigeration, autos," or, as he adds, "twin beds" (*SE* 176). Williams thus explores the implications of a machine aesthetic, suggesting again that although the geometric or machine style in art and architecture was inspired in part by the functional look of technological products, it is important to separate the aesthetic from the everyday objects that embody it.[26]

Henry Sayre has argued that Williams's definition of a poem as a machine made of words may draw on *Towards a New Architecture* (1927) by the French architect, Le Corbusier, who defined a house as "a machine for living in."[27] Whether he read *Towards a New Architecture* or not, Williams would have known that modern architects celebrated the same machine aesthetic he found attractive. His essay implicitly cautions that, especially in architecture, modernity might falsely appear to consist of functional external trappings. He points to the breakdown of the analogy between the uses of art and the uses of technology: Art, he notes, does not separate people (in twin beds) or lead to a lack of contact. The argument, found also in the Jacataqua section of *In the American Grain*, is that although art possesses many of the virtues of technology and science, it is not implicated in the production of things such as bathrooms, kitchens, or hospitals, which are denounced as bars "to more intimate shocks" (*IAG* 177), that is, as ways of avoiding contact.

The same passage in *In the American Grain* identifies poets as having a healthy outlook: "Poets? Where? They are the test" (*IAG* 178–79). The implication is that poets have a sense of immediacy that does not look to utility or results, in contrast to what Williams describes as the usual American way of life that "drives us apart and forces us upon science and invention—away from touch" (*IAG* 179). In short, the uselessness of art becomes its strength. Similarly, in the 1937 essay, Williams argues that architects and poets may be like engineers, but they are not primarily practical. As Le Corbusier wrote in *Towards a New Architecture*, anyone who can say only "my house is practical," will say to the architect, "you have not touched my heart."[28]

Williams at times shifts from discussing engineers to discussing more theoretical scientists, and he again uses the analogy between scientific and artistic creativity to place poetry on firmer ground. His 1918 "Prologue" to *Kora in Hell* rejects "a science doing slavey service upon gas engines" in favor of "inventive imagination" and "the authentic spirit of change" (*I* 13), which do not rest content with any particular invention. Similarly, although Pasteur was often cited as the prototype of a scientist who produced socially important results, in 1951 Williams described him as interested in "theory" (*SSA* 62), not results.

Having reassessed the strengths of science, Williams can insist upon the superiority of poetry. In *The Embodiment of Knowledge*, he compares Shakespeare and Bacon: both men are creative, but Shakespeare is used to dramatize the fact that creativity is more important than its results. As another of Williams's essays explains. Shakespeare makes "objects, realities which he has to abandon to make another, and another—perfectly blank to him as soon as they are completed" (*SE* 56). Also in *The Embodiment of Knowledge*, Williams discusses machinery (meaning literal machines but also poems and knowledge in general) in similar terms. "As soon as we make it we must at once plan to escape—and escape" (*EK* 62), adding that the problem with the engineer is that his knowledge "being so in particular, is likely to absorb him into itself until he becomes a scientist [here, obviously, a technocrat]—limited, segregated—unable to escape" (*EK* 63). It is not surprising then that the next section of *The Embodiment of Knowledge* compares Bacon unfavorably to Shakespeare. As Emily Wallace has pointed out, Williams saw the "artist's strategy of escape" as a way of affirming diversity and resisting any closed systems, political, social, commercial, or artistic.[29] For Williams, one strength of the poet was that his knowledge was not as easily harnessed for narrow ends as that of the engineer or scientist.

Poetic creativity, then, is described as similar to but purer than scientific creativity. Yet in the American context this position too presented problems, since at times it was not only the style and creativity but the public acceptance of technology that Williams wanted to claim for poetry, even as he challenged the reasons why Americans valued science and technology. In a 1917 letter to Harriet Monroe, Williams clearly articulated his frustration at the lack of a public forum for poetry. He told Monroe that she should stop paying the poets whose work she published because she was endangering the continued financial security of her magazine, *Poetry*: "You are jeopardizing the existence of your magazine in the mistaken notion that what poets want is money, when in reality—though money is sorely needed also—they need space. . . . This lack of space, this lack of opportunity to appear is the hell" (*SL* 41).

The preface to a poem published in the summer of 1921 in Williams's magazine, *Contact*, is more ironic and wittier, but it equally reveals Williams's dismay about the lack of an audience for poetry in modern, which is to say technological and commercial, society. He presented his poem as a commercial sample, and offered to sell poems with "those of most excellence, as in all commercial exchange, being rated higher in price."[30] The gauge of excellence, he added, would be "length and success" (p. 2). Obviously, Williams wanted his readers to conclude that poetry could not be measured in the same way that commercial products could be. And yet the irony has a bitter edge, for part of Williams, like Dr. Paterson, "envie[d] the men that ran / and could run off / toward the peripheries— / to other centers, direct— / for . . . . authority in the world" (*Pat* 36).

Williams never explicitly articulated the problems involved in attempts to grant authority to poetry by comparing it with science and technology. His writing, nevertheless, frequently acknowledges such problems and consistently confronts the American tendency to equate science and technology with practical results or actual commodities, this being the tendency that made metaphorical appeals to science and technology both attractive and problematic.

In *The Embodiment of Knowledge*, written between 1928 and 1930, for example, Williams argues that Americans must be reeducated, and he strikes out against science, technology, and philosophy, including educational philosophy, because they all, at least as popularly understood in America, treat knowledge as a commodity separated from everyday living rather than as something informing people's lives.[31] The "partial and tentative" knowledge of poetry (*EK* 6) is specifically set against the dogmas of science and philosophy. Williams writes: "John Dewey and others appear to look for a solution to the problem of education in psychology and sociology—in philosophy then. They might do worse than to seek it in poetry. It is the poetic conception (see Einstein's reported statement—among others) of the universe that is the correct one" (*EK* 7).

Williams's critique of philosophy is complicated, since he sets philosophy, which reaches after fixed solutions, against more flexible ways of thinking. The writer's ideas, he continues, come "in the act of writing" (*EK* 7). Yet Williams also argues that poetry is scientifically validated as "correct," which seems to contradict both his point that one should not look to science or philosophy for answers and his insistence that knowledge should be an activity rather than a matter of reaching final answers. Williams's understanding of Einstein, however, suggested that a *correct* conception of the universe was not a fixed conception. Williams's meditations on what education should be, moreover, allow for contradictions, as philosophy in his view does not. In fact, although Williams's desire to grant poetry authority is heartfelt, his appeal to scientific correctness is by no means his final word on the subject. His next paragraph reopens the question by suggesting that poetry might also learn from philosophy, but that "the courtesy," which in a sense Williams has just shown in his appeal to Einstein, should be reciprocal (*EK* 7).

In many ways *The Embodiment of Knowledge* repeats arguments Williams had already made five years earlier in the 1923 *Spring and All*, which also attacks American science and education insofar as they represent "dead dissections" (*I* 138).[32] As his treatment of Dewey in *The Embodiment of Knowledge* suggests, Williams had some reservations about Dewey's position, yet he also agreed with him on many points. In a 1918 article entitled "Education and Social Direction," for example, Dewey stresses the interdependence of ends and means in education, arguing against "unintelligent convention, unexamined tradition [at the expense of] personal originality."[33] An article that appeared one year earlier, "Current Tendencies in Education," concluded: "The mobility and flexibility, the freshness and variety, of modern life stand in increasingly startling contrast with the wooden routine, the deadly conventionality, of the average traditional school."[34]

Dewey, like Williams, also attempts to rescue culture from those who see knowledge as divorced from the everyday, modern world. In "Education and Social Direction," Dewey struggles to demonstrate that a liberal education, including a study of the arts, is necessary in a democratic society.[35] Without wanting to jettison facts or particular skills, Dewey champions thought. As he says most concisely in the 1916 *Democracy and Education*, which anticipates the points made in the articles: "Where there is reflection there is suspense . . . Since the situation in which thinking occurs is a doubtful one, thinking is a process of inquiry, of looking into things, of investigating. Acquiring is always secondary, and instrumental to the act of *in*quiring."[36]

Many of Williams's works are in Dewey's sense acts of thought. Suspense may be too calm a description of Williams's strategy, especially in the prose, although he himself uses the word in *Spring and All* (*I* 120). In any case, it is in this manner that Williams's work continually dramatizes and explores rather than announces solutions to the problems he confronted as a poet in America. Not surprisingly, then, Williams's questions about how to view America's technological and scientific strengths change depending on his audience.

In *Spring and All* and *The Embodiment of Knowledge*, Williams's argument is with those who champion science, technology, and philosophy in order to treat knowledge as a commodity, as a matter of acquiring rather than of inquiring, to use Dewey's distinction. Elsewhere, Williams defends science against those skeptical about whether Americans could be creative. Whatever his audience, however, Williams's ambivalence about science and technology remains.

In *In the American Grain*, Williams contends, for example, that the schemes like the actual products of scientists, saints, and philosophers "amount to nothing. But to the men themselves every moment, every detail, the devotion, the clarities are vital—and so we value them, in short, by their style" (*IAG* 207). Williams here celebrates scientists with a familiar shift: scientists, saints, and philosophers (like the inventors of American technological products) are important because of their creativity, which involved a certain mode of operation including a detailed attention to particulars. It is how they work, not the results of their work, that is important. Williams, however, debunks the detailed, focused method of proceeding (that is, the scientific method) when it is practiced by men of science rather than by poets. Franklin's "itch to serve science" and "to do the little concrete thing" (*IAG* 155) is found lacking, as is the scientists' urge to keep things "cold and small and under the cold lens" (*IAG* 175). If Williams celebrates the strengths of American inventors, he nonetheless insists that by paying too much attention to the practical results of their work, scientists deny the creative force of their own methods.

In his praise of scientific creativity, Williams also invokes the material success of technology and applied science. His use of the word *style*, for example, blurs the distinction between process and product. That is, his description of the importance of detail and clarity to creative men is, by metonymy, a description of the style of technological products or inventions, and so of the machine aesthetic that attracted him. In the ambiguity of the word style, Williams's rhetoric subtly dramatizes the way literary and artistic style in his day was evoking the look of commodities and the value granted to scientists or engineers because of the practical results of their work, not because of the creativity poets and scientists or engineers shared. By first inviting and then explicitly denying a misreading of *style* as the style of a manufactured object rather than as a method of operating, Williams draws on typical American attitudes toward scientists. In short, even in his praise of scientific creativity, Williams shows his sensitivity to the ways in which creativity and results were identified in an American context.

Indeed, the aura of importance surrounding science, precisely because of its practical results, is something Williams claims for *himself* in his late self description as one trained with "the humility and caution of the scientist" (A 58). Especially as a doctor, Williams felt the force of the American attitudes he challenged, and he could clearly see that poets did not achieve the same kind of results that doctors did. As he has a woman say to the doctor-hero of A *Novelette* (1930): "It was damned clever, making a diagnosis like that and saving a baby's life, worth more than any poem, I think" (*I* 279).

In his early writing, Williams does not resolve the problem of how to define and defend American poetry. Yet his attempts to appropriate the popular success of technology along with its aesthetic or to redefine science and technology so as to insist on the value of poetry constantly acknowledge the tensions caused by the use of analogies between science, technology, and poetry. In particular, Williams's increasing awareness of how the style of some of his poems might be unfavorably compared with the style of commercial products by the very American audience he wished to reach appears to have fed his desire to forge a new description and a new defense of his poetry.

As Tichi has shown, in Williams's world, in its advertisements, medical texts, and urban landscapes, "mechanistic thinking [was] . . . recognized truth . . . not open to self-consciousness, much less to challenge."[37] Given also the visual arts' attention to the mechanical, Williams's admiration for a machine aesthetic is in ways, as Tichi says, unremarkable. What is remarkable is how Williams in his prose tacitly but repeatedly confronted the problems raised by modernist defenses of a machine aesthetic in America.

The problems Williams considered did not solely concern poetry's reception. The relationship between a writer's poetry and poetics is one of reciprocity. That is, a writer's ideas about poetry are not simply a prescription for or a case study of his practice, but involve constant engagement between poetry and ideas. To demonstrate the relationship between Williams's defenses of poetry and his own poems, I should like to offer a reading, in accordance with Williams's stated poetics, of one poem in which he relies on the modern style related to a machine aesthetic, and then to comment further on the implications of the usual modernist strategy for defending such poetry. My point is not to suggest that the poem itself is in any way diminished by the poetics that might be used to describe it, but rather to show why Williams felt the need for a new way to describe what he was doing.

"Young Sycamore" (1927) is a brief poem of one sentence that opens:

> I must tell you
> this young tree
> whose round and firm trunk
> between the wet
>
> pavement and the gutter
> (where water
> is trickling) rises
> bodily
>
> [*CEP* 332]

The poem follows the tree's rise to a fork, and then to smaller branches with cocoons in them, "till nothing is left of it / but two / eccentric knotted / twigs / bending forward / hornlike at the top" (*CEP* 332). The poem is based on an aesthetic shared with paintings and photographs by members of the Stieglitz circle; indeed, Bram Dijkstra argues that Williams is literally describing Alfred Stieglitz's photograph, *Spring Showers*.[38]

"Young Sycamore" and other Williams poems like it seem to resist explication, to be purely descriptive, although, as J. Hillis Miller points out, Williams takes a firm stance against "the falseness of attempting to 'copy' nature" (*I* 107), desiring "not 'realism' but reality itself" (*I* 117). Miller's argument is that the poem does not represent a tree, but rather as a poem it "is an object which has the same kind of life as the tree."[39] If, as he argues, "Young Sycamore" is not symbolic but an object in its own right, it nonetheless presents an analogy between the growth of a tree and the growth of a poem. The motion described is paradigmatic, familiar to any reader of *Kora in Hell*,

Spring and All, or *In the American Grain*.[40] The tree's "bodily" rise ending in the near still life of the two twigs is a motion very like Williams's description of imaginative creation.[41] In "How to Write" from 1936, Williams says that poems begin with "the very muscles and bones of the body itself speaking, [although] once the writing is on the paper it becomes an object. . . . an object for the liveliest attention that the full mind can give it" (*SSA* 98). The poem takes its place as another artifact, an object in the world, but also refers to a series of parallel motions: Nature produces the tree, in a fashion very close to the way we have seen the production of inventions described, Williams produces his poem; and the reader is invited to join in the creative process, not by looking through the language to that which it describes, but by paying attention to the poem itself, and, if the paradigm of the poet and nature holds, producing some object of his own.[42]

Even without knowing Williams's theories about poetry, by turning mind and attention to the poem-as-object, the reader is referred to Williams's process of creation in language. Although the poem describes an act of detailed perception, and thus at first recalls Williams's statement that artists teach us to see,[43] a closer examination of "Young Sycamore" shows that it places equal emphasis on speaking and language. The urgency of the first line focuses attention on the poet's voice, while the careful syntax of the main clause and the enjambment suggest that Williams is not showing us a tree or even clearly telling us about one; he is creating a tree of language: "I must tell you / this young tree."

At the same time, it is no accident that the verbs, like the tree, thin out towards the end of the poem, nor that the one simile occurs in the last line, as the flow of language ends. Williams said he disliked similes: "The coining of similes is a pastime of very low order" (*I* 18). It has been a commonplace of Williams scholarship to use such statements as evidence of Williams's desire to present poems as objects or as being objective descriptions of discrete, nonsymbolic objects in the world.[44] As Henry Sayre argues, however, Williams's similes and metaphors are not lapses by someone who wished to but could not avoid "subjective observation and interpretation"; they serve rather to make the poems work as "the site of the interplay between the mind and reality."[45] The overt figure at the end of "Young Sycamore" deliberately calls attention both to the poet's act of interpretation and to his linguistic creation. It shifts our attention from trees to poems at the same time that it recasts the poem as an emblem or, at least, as a series of figures. Not only is the tree's growth like the poet's creative process, but the waxing and waning of the tree is also echoed more succinctly in the cocoons' simultaneous image of death and potential life, and in the contradictory motions finally abstracted in the stark image of the two knotted twigs.[46]

The central concern of the poem is the process of growth or creation with its inevitable culmination in an object. But objects can yield new realities, new life, like the cocoons or like the emblem of the eccentric twigs. The twigs are eccentric because they are off center, leaning forward,[47] but also because the literal center of the poem images the process of creation yielding both multiplicity and destruction as the trunk's "one undulant / thrust" begins to divide between the third and fourth stanzas, and the poem widens its focus to include the cocoons, which, as a gardener like Williams would have known, spell destruction for trees. The final image is eccentric as well because the twigs remain stubbornly particular, even as they tempt us to align their double nature with the other dualities to which the poem calls our attention—the process behind the product; the cocoons' destruction in creation; the poem as object and figure, physical and intellectual. As Williams says, "nothing is left of it / but two." Delaying the noun by a stanza break and a line of adjectives, Williams makes "two" seem for a moment the object itself. The final duality is formal or structural.

One might say that all of the oppositions suggested in the poem are given their purest expression—not resolved, but expressed—in the image, which is the fruition of the poem, but which is so well "knotted" that nothing remains to be said: "The detail is its own solution." The twigs for example, suggest both an upward movement (*top* is the final word of the poem) and a return to earth in their horn-like bend. Like the ornament and the steeple or the contrast between a squat edifice and the moon in "To a Solitary Disciple" (*CEP* 167–68), the image of the two twigs moves Williams's readers in two directions simultaneously. Similarly, the poem both is and is about a unique object even as it suggests that all objects, once subjected to lively attention, can be made to release a creative energy that necessarily transcends the discrete structure of individual objects.[48]

Finally, the result of the creative energy alluded to in the poem is the revelation of a structure. As the life of the poem is parallel to the life of the tree, creativity is itself another version of the structure imaged by the entire movement of the poem from the poet's presence, insisted upon in the first line, which roots the poem in a creative, human, speaker, to the final equation of the formal essence of the poem and the bare architecture of the leafless tree.

Williams often described the force or energy of poems as an essence or rare presence (A 362). These images occur quite early in his writing. A 1921 editorial in *Contact*, discussing Burke's article on Laforgue, describes the search for a "milligram of radium" (*SE* 36), while the essay on Marianne Moore published in *A Novelette and Other Prose* speaks of the "white light that is the background of all good work" (*SE* 122). In the Rasles section of *In the American Grain*, Williams isolates within himself a "core of nature" (*IAG* 105) analogous to "the strange phosphorus of the life" (*IAG* v) that he seeks throughout the book. At all stages of his career, Williams refers to an incandescent universal presence that informs and perhaps is in all art and all natural objects, but the status of this essence varies. In "Young Sycamore," at least, it is presented as a structure, and finally as impersonal, and this reading is reinforced by the poem's use of language. That is, the visual regularity of the quatrains and the sparse, unembellished lines and vocabulary, like the imagistic progression of the poem, add to this effect.[49]

Descriptions of such poetry as impersonal, or at least as objective, are commonplace in modernist poetics, and the link between this style of poetry and science is widely noted.[50] Moore identified Louise Bogan's terse pronouncements as being "rendered with laboratory detachment" (*R* 230). And Eliot, to choose a poet with whom Williams generally took issue, suggested that it is in "depersonalization that art may be said to approach the condition of science."[51] Williams's own descriptions of the use of language in poems such as "Young Sycamore" similarly invoke science, as when he approvingly says in his 1931 review of Moore's poetry that words are "separated out by science, treated with acid to remove the smudges, washed, dried and placed right side up on a clean surface" (*SE* 128).[52] Here Williams praises removing the emotional and conventional associations of words, and sees the poet's cleansing activity as akin to the laboratory or assembly line worker's.[53]

Williams thus recognizes that his insistence upon words as objects or structures, central to the style and content of "Young Sycamore," might be linked to science, to technology, and ultimately to the products of technology. The link, indeed, is subtle. The poem not only appeals to a taste for the clean lines and efficiency associated with industrial technology, but it is defined as a discrete structure, a "machine made of words" that does not need to refer outside of itself for its effect since "its movement is intrinsic" (*SE* 256).

Williams describes the need to reclaim the essence of poems, trees, and other objects "nameless under an old misappellation" (*IAG* v) with a new, cleaner language to be provided by the poet. What the poet reclaims then, in "Young Sycamore," is not a particular tree, but a method of reclamation and an essential structure that is in one aspect purely formal. In fact, "Young Sycamore" 's reference to the poet's process of creation recasts creativity itself not only within but also as a formal structure.

To force Williams's poetic to one possible conclusion, it is not that poems, more self-consciously than machines, reveal human inventiveness, but that knowledge, language, poems, plants, and men are structurally similar, and their structural essence is best described by analogy to machines or technological products. Indeed, in 1919 Williams wrote in an article for the magazine, *Others*, with which he was involved: "Poets have written of the big leaves and the little leaves, leaves that are red, green, yellow and the one thing they have never seen about a leaf is that it is a little engine. It is one of the things that make a plant GO."[54] Similarly in "Young Sycamore," the human mind, another example of a biological design, is reenvisioned as a mechanism, analogous to the formal structure of the poem.[55]

Williams, however, was not fully comfortable with this view to which his acceptance of a certain style seemed to commit him.[56] In the 1937 dialogue on poetry and architecture, for example, he tries to explain why he rejects the " 'back to humanity, back to the soil' business" about the organic production of art while still believing that people are the "origin of every bit of life that can possibly inhabit any structure" (*SE* 178). His prose reflects his difficulties with the vocabulary he had available to him. People, he continues, "represent, in themselves, the structure which art . . . Put it this way: If we don't cling to the warmth which breathes into a house or a poem alike from human need . . . the whole matter has nothing to hold it together and becomes structurally weak" (*SE* 178).

The first ellipsis in the above quotation is Williams's, and seems to indicate his unwillingness or inability to say what the structure of art has to do with the structure of people. The logical way to complete the sentence would be to suggest, as "Young Sycamore" suggests, that art also represents, or repeats, the structure found in human beings. Such a conclusion, however, does not locate people as the origin of the life that inhabits structures like plants as well as poems and buildings. And Williams usually wanted to insist, as an essay from around 1926 entitled "What is the Use of Poetry" put it, that poetry "returns authority to man."[57] Hence he stops and proposes instead that all inventions arise from human need. This proposal avoids the more radical implications of adopting a machine aesthetic, but still it does not fully answer the question of how, in the practical American context Williams set for himself, one might show that poetry is important and necessary. Indeed, moments such as this in Williams's prose underline why he felt the need to find a more convincing way of defining poetry and its importance.

In 1944, Williams cautioned that the "arts have a *complex* relation to society" (*SE* 256). Exploring one aspect of his attempts to define this relationship illuminates the development of Williams's poetry and poetics. More importantly, in adopting a modern style commonly associated with the rise of technology and in simultaneously attempting an analysis of American modernity generally, Williams reveals the difficulties involved in sustaining a defense of modern poetry or in describing its importance given the values of modern America and the Americans for and to whom Williams wanted his poetry to speak.

William Carlos Williams: A New Way of Measuring

Williams described his first attraction to a machine aesthetic as due in part to his desire to be allied with avant-garde movements in a number of the arts. The same desire may initially have fed his willingness to experiment with other styles and theories associated with modernism, some of which coexist uneasily with the style and poetics just discussed.[58] Williams's attempts to outline the nature and importance of a more loosely structured poetry also cast light upon the endeavors of American modernists in a number of fields, and upon the sometimes contradictory poetics identified as modernist. Further, Williams's thought reveals the importance of the relationship between science, technology, and the arts to the American modernist project.

The more loosely structured style, which Williams says he "liked to do most of all,"[59] is evidenced in *Kora in Hell,* as well as in the later "The Sea-Elephant" and "The Trees," poems that differ markedly from poems like "Young Sycamore." The analogues proposed for this looser style, both by Williams and by his critics, are many, including Dadaist and Surrealist experiments in the plastic arts, Bergson's philosophy, and the creative and natural energy at the root of scientific advances, an energy that in "Young Sycamore" was defined structurally.[60]

Williams's belief that poetry reveals a universal presence remained constant, but his characterization of the nature of that presence changed, both explicitly and implicitly, in his defense of what may be called a fluid style.[61] When he calls for "the authentic spirit of change" (*I* 13) or for a "loosen[ing of] the attention" (*I* 14) in *Kora in Hell,* the style of that poem, almost a stream of consciousness, calls attention to a verbal energy, which one is invited to see as an expression of inner character or of the mind at work. This characterization of imaginative energy does not insist upon the hard-edged, sparse style that Williams elsewhere defended. Indeed, in 1938 he said of Sheeler's later Precisionist paintings, "someone should smash his camera and open his brain," a charge that does not accord with Williams's more frequently and publicly voiced admiration for the artist.[62] The 1938 outburst is best explained by Williams's struggles in the 1930s with his own attraction to a machine style that he felt might not adequately emphasize, or accurately characterize, the creative activity he so often celebrated and identified as central to his defense of modern poetry.

In the group of poems entitled "Della Primavera Transportata Al

Morale," Williams associates mental activity with a loosened style and with the images of radiance, heat, and light in terms of which he frequently discussed "the radiant gist" (*Pat* 186) revealed by or released by poetry. In the title poem ("Della Primavera Trasportata [*sic*] Al Morale"), Williams offers a series of morals:

> Moral
>
> it looses me
>
> Moral
>
> it supports me
>
> Moral
>
> it has never ceased
> to flow
>
> [*CEP* 59]

The reference of *it* is left ambiguous; while *it* could refer specifically to the confluence of sights and sounds (a truck's clatter, the wind, a river) described just before the morals are introduced, *it* also, more generally, refers to the whole creative outburst associated with April and beginnings in the poem's first lines. The very multiplicity of reference underlines Williams's association between creativity and fluidity or looseness.

The poem concludes in a familiar vein, announcing that the "forms / of the emotions are crystalline, / geometric-faceted"; however, the style of the poem mimics the "white heat of / understanding, when a flame / runs through the gap made / by learning" (*CEP* 64). The gap, presumably, is similar to Stevens's "dumbfoundering abyss / Between us and the object" (*CP* 437), a gap Williams fills between crystalline objects and crystalline emotions by the flame of the creative imagination in motion.[63] One of the most memorable features of the poem is, in fact, its style. There is no consistent stanzaic form maintained; public signs, a menu, and literal images are juxtaposed with a more personal, even colloquial, voice. These juxtapositions are, in effect, as important as the individual images or voices. The poem can thus be seen from one point of view as an engineering feat, yet the image Williams provides for creative process here is the image of a flame. Moreover, the rapidity with which images are presented and replaced, the lack of any obvious formal structure, and even the use of ambiguous pronouns (such as "it") deliberately present the poem as if it was composed in a "white heat." The poem thereby makes us focus upon fluid speaking or motion—a figure for the energizing relationship between observer and observed—as the basic characteristic of creativity.[64]

Mike Weaver describes Williams's growing concern with motion in the 1930s and convincingly argues that Williams draws on current discussions of relativity and, from 1927, on Whitehead's *Science and the Modern World*, in formulating a new poetics.[65] The new science, particularly Einsteinian physics, played an important role in Williams's defense of a fluid style and, one might add, in Williams's eventual reconciliation of his two styles in his mature poetics. The way in which Williams was influenced by science was not straightforward, however.

For one thing, Williams's concern with motion, as seen in the style of *Kora in Hell*, predated his exposure to the new science and, early on, owed as much to the American fascination with speed, to the Surrealists, and to modernist poetics generally as it did to Whitehead.[66] It was not until 1927 that Williams's discussions with John Riordan provided him with a detailed introduction to the new physics, and it was not until some months later that Williams, reading Whitehead, pronounced *Science and the Modern World* a "milestone" in his career.[67] His statement suggests that Williams had not previously seen exactly how the new physics might be relevant to his poetics, although by 1946 Whitehead would join Dewey and the Surrealists as one of the saints of Williams's late marriage of process and structure in "Choral: the Pink Church" (*CLP* 160–61).

Einstein and the Curies do appear in Williams's early writing, although his early images of Einstein and of motion draw mostly on the common misconceptions about the relationship between Einsteinian physics and Bergsonian philosophy. Given the popular conflation of science and technology, Williams's early view of Einstein is better seen in light of his general ambivalence about science and technology than in light of Einstein's actual theories.

Still, examining the image of Einstein in Williams's writing before 1927 does yield interesting results. Einstein appeared in early issues of *Contact*, and in the 1921 "St. Francis Einstein of the Daffodils,"[68] which prefigures both *In the American Grain*, and, more importantly for the point I am making about the continuity of Williams's experiments with a fluid style, "Della Primavera Trasportata Al Morale." Not only are images of a new beauty and spring juxtaposed with images of urban slums and American commercialism in both poems, but the "topbranches / [swaying] with contrary motions" of "St. Francis Einstein" become "the tree moving diversely / in all parts" (*CEP* 60) in the later poem. Both poems set the eccentric twigs of "Young Sycamore" in motion, imagistically and stylistically, and both suggest that style will be an important part of Williams's portrayal of motion or underlying energy.

In 1921, Williams had Einstein preside over his fluid style in part because of the connection popularly drawn between relativity theory and a universe of flux. Einstein was further called upon to defend the artists' mental flux, or the constant dismantling of old forms in order to create anew, which was commonly associated with a fluid style. The extension of the idea of relativity to the inner as well as the outer world also identified Einstein's constant, the speed of light, with a dynamic motion, which Williams had already imaged as light or flame in motion. In other words, already existing notions of an inner motion as a universal human essence came to be associated with Einstein.

Williams's early use of Einstein, then, sought a scientific foundation for a poetics that he and others had begun to develop well before Einstein's public appearance. Einstein's primary characteristic for Williams in 1921 may have been that he offered a new vision of the world, just as Williams wanted to do in his poetry. Further, at least in 1921, the uses of the arts and of relativity theory were more easily compared than the uses of poetry and of bathroom or kitchen fixtures. Einstein seemed to offer a definition of the modern that, along with his popularity and the scientific *truth* of his theories, might be appropriated for poetry. Moreover, the particular popularity of Einsteinian physics in America seemed to offer proof that Americans could admire and find a place for a creative intelligence, the results of which were not measured commercially.

For Williams, constant creation was associated specifically with the production of language in poetry. Thus, in the January 1921 issue of *Contact*, St. Francis (a figure dubbed St. Francis Einstein by the summer issue) is the saint of communication. St. Francis, says Williams, found "a common stem where all were one and from which every paired characteristic branched" (*SE* 28). Here, as in "Della Primavera Trasportata Al Morale," creative activity is emphasized and imaged as Einstein's constant moving freely from object to object, providing the medium by which all things are related. The image of "a common stem . . . from which every paired characteristic branched" also recalls the image of the horned twigs in "Young Sycamore." Einstein, however, provides a way to reenvision creativity in an image that maintains Williams's insistence on structure without invoking a technological metaphor to describe creative process.

Nor is it accidental that Williams associates Einstein with St. Francis. A somewhat secularized St. Francis was a familiar figure in America, representing the values of simplicity, self-control, and work to a middle class in need of moral models.[69] By recruiting St. Francis, Williams underlined the commonly voiced perception that modern culture was not fulfilling human needs. He drew also on America's (and his own) warm response to Einstein the man, whose public monument in Washington even now portrays him as someone into whose lap children are invited.[70] The image of St. Francis Einstein thus grows out of Williams's hope that the enthusiasm with which both Einstein and St. Francis were adopted by America might be turned toward poetry, which could offer Americans the immediate and imaginative knowledge of the world that their admiration of the scientist and saint suggested they wanted. The desire to understand one's world, Williams wrote, "in America, can only be filled by . . . a poetic knowledge" (*IAG* 213).

In short, Einstein provided the image Williams needed to define a place for modernist art in American society, offering a different and in ways less problematic analogy than the comparison between technology and art. At the same time, Williams was wary of allowing scientific truths to underwrite his poetry. Despite his and others' equation of science and poetry, Williams was sensitive to the fact that the papers were not inclined to grant visiting artists the same enthusiastic reception that they granted Einstein. Williams thus wavered between claiming Einstein for poetry, claiming poetry's superiority to science, and, with some justice, casting doubt on the reasons for Einstein's status as a celebrity.

One can see Williams's defensiveness about Einstein best in his essays. As Carol Donley has shown, the press first provided the suggestion, echoed by Williams, that Einstein was merely a follower, a latecomer as a "Poet in Science."[71] In a 1934 article, Williams says that scientists "haven't understood that one plus one plus one plus one plus one equals not five but one," and yet hastens to add that "even science is beginning at last to catch on to it", that is, science will validate poetry, but poetry first discovered scientific truths.[72] To return to Williams's statements in 1921, St. Francis Einstein is the saint of communication and of process, "the patron saint of the United States" (*SE* 27), but sainthood has a questionable status. "A patron saint is one thing but in the intercommunications of art there should be something more" (*SE* 27). Poetry, Williams goes on to suggest, will not be equivalent to science but might be more relevant to society.

Given Williams's ambivalence about science in the essays, it is not surprising that his first poem about Einstein displays the same mixed tone. In "St. Francis Einstein of the Daffodils," one finds Einstein associated with the cycles of change that originated in *Kora in Hell.* Einstein—a familiar, if not easily visualized, combination of Columbus, Venus, and Spring—arrives in the new world and frees it from classical notions of beauty, from the past. He brings both creation and destruction, presiding over an old negro who invites cats (including Lesbia's black cat) to eat poisoned fish. The juxtapositions are, initially, characteristic of Williams's insistence on change: southerly winds follow northerly gales, creation follows destruction, to leave us with the motions of awakening and uncovering.

Although the poem's style reinforces our sense of motion and fluidity, Williams mockingly presents it as his newest product, prefacing what he calls a sample poem with the announcement that a "minimum price of fifty dollars will be charged for all poems" (p. 2). This introduction to "St. Francis Einstein of the Daffodils" apparently was called forth by Einstein's success in America; and in the poem itself one finds a similar tone. Einstein arrives "to buy freedom / for the daffodils" (p. 2), and he arrives "in fashion" as the "wise newspapers / . . . quote the great mathematician" (pp. 2–3). The irony is evident, as is Williams's uneasiness with Einstein's popularity, which he links with the commercial success of technology, even as he claims Einstein's theories for poetry.

Similarly, in *The Embodiment of Knowledge* Williams has a somewhat confusing set of notes on the European modernists' successful predictions of war, followed by some comments on Lewis's criticism of Pound. Williams ends his discussion by saying that "we are convinced by this pander because, by God, what he says is scientifically . . . true. We buy his book" (*EK* 84). What Williams thinks Pound's book, or Lewis's criticisms of physics and Pound, have to do with science, or with the accuracy of literary predictions, is unclear. What is clear is that the image of prostituting oneself is tied to the commercial and popular success of those books that can be called scientifically true. While Williams wished to show that poetry was as *true* and as important as science, he wished also to rescue poetry from competing with science for the truth. This was in part due to his continued sense that scientific truths could have commercial value, while poetic truths did not.

In the course of his rescue mission, Williams also offered a plea that theoretical science not be made subservient to technological or commercial demands. "St. Francis Einstein of the Daffodils" begins Williams's serious critique of the temptations to which theoretical science is prey, a critique developed more fully in *In the American Grain*. Einstein, in the early poem, positively affirms the necessity of theoretical intelligence, although he shares with art and with America generally the destruction entailed by true creativity. The poisonous fish-heads said to be described by Einstein's mathematics are an image that Williams elsewhere associates with necessary destruction. In *In the American Grain*, he has Columbus (a figure for Einstein in "St. Francis Einstein") discover diverse trees, strange fish (*IAG* 26), and an "acrid and poisonous apple' (*IAG* 7). Although neither *In the American Grain* nor "St. Francis Einstein" uncovers a snake in the garden, a late essay on Sandburg identifies theoretical intelligence as "the active worm in the fruit" (*SE* 273), mixing the images of snakes, worms, and apples. All images of the necessary evil in paradise are associated with the destruction that Williams throughout his writing insists must accompany creation in order to make room for yet newer creations.

In this context, it is significant that Williams calls both scientists and saints snake charmers (*IAG* 207), and that he remarks on how the "mysterious is so simple when revealed by science," after quoting from a prosaic scientific explanation of how the "fiery serpent that bit the children of Israel when they wandered through the wilderness was possibly the guinea worm" (*IAG* 222–23). Williams's point seems to be that the creative intelligence of scientists (snake charmers) is too easily domesticated, as with Einstein, too prone to accept commercial and popular support for its products and thus to accept easy answers.[73] Certainly this is Williams's point when, later, he asserts that the "true nature of the revolutionary St. Francis—whom they [the puppet-makers] recaptured and subjected to their rule"—is the nature of "the running animal" (*SE* 246). That is, neither saints nor scientists (nor, implicitly, poets) should serve institutional or narrowly partisan goals; their virtue lies in the example set by their mental and imaginative agility; like Shakespeare in *The Embodiment of Knowledge*, they are meant to escape closed systems.

It is worth pausing to note that Einstein's importance to Williams's modernist poetics, like the importance of contemporary technological developments, grew out of the need shown in Williams's early writing to discover what the arts have to offer in an age dominated by science and technology. The analogies between poetry and physics, then, like the analogies between poetry and technology, are not simply reflections of specific contemporary developments so much as attempts to solve the general problem of how the arts might be related to, without being subsumed by, the sciences. Williams applauds the validity and the respect granted to the *true* Einstein, criticizing only the domestication of science. At the same time, he insists that poetry already has the virtues but not the vices of science, ignoring his own sometimes envious recognition elsewhere that poetry's temptations are not those of science.

Williams's use of Einstein's ideas changed after his exposure to slightly more detailed versions of the new science, as can be seen in his descriptions of the variable foot in poetry and his related use of the Curies' work on radium.[74] Einstein, specifically, continued to be of importance to Williams because his theories, as Williams came to understand them, served to reconcile the poetics of structure with the poetics of process and thus helped Williams forge a unified poetics and a stronger defense of poetry.

Before offering a brief sketch of how Williams used Einsteinian physics to reconcile his two styles, let me add that Williams's ambivalence as to whether science underlined the importance of modern poetry or eclipsed poetry persisted throughout his career. Thus, any consideration of Williams's later use of Einstein needs to take into account why Einstein was important to Williams and why Williams was so nervous about claiming Einstein. In his notes for an address at the University of Washington, given in 1950, for example, Williams proclaimed himself "just a literary guy, not *practical*—like a one-time atomic physicist," adding, "What a nerve to come to a going institution of learning to teach you how to write?! Even to sell? Why, you might as well have an Einstein. HE at least can play the violin" (*SE* 298). The dash makes the syntax here confusing. Is Einstein more practical than Williams, and a commercial success for the wrong reasons? Or is he comparable to Williams ("just a literary guy, not practical") with only the slim advantage of being known to play the violin, as Williams withholds the fact that he too was an amateur violinist? The ambiguous image of the physicist-musician leads the reader to look past the negative image of an overly successful scientist to see the "ultimate Einstein," who is really Williams, claiming for himself the validity and appeal he too found in Einstein.

Moreover, after Hiroshima, Williams wrote not only of the Curies' and Einstein's pure creativity, but also of the power, both literal and conceptual, of the technology made possible by the new scientific breakthroughs. "Asphodel, That Greeny Flower" self-consciously admits that the atom bomb is something against which it is difficult to pit poetry: "We come to our deaths / in silence. / The bomb speaks" (*PB* 168). And yet Williams ends by affirming man's ability to geld the bomb (*PB* 179) by his imaginative perception of it:

> . . . Don't think
> that because I say this
> in a poem
> it can be treated lightly

or that the facts will not uphold it.
Are facts not flowers
and flowers facts
or poems flowers
or all works of the imagination,
interchangeable?

[*PB* 178]

"Only the imagination is real!" (*PB* 179), we are finally told; however, the poem acknowledges that a metaphorical appropriation of the results of the new physics will not solve the world's problems.

Although there is a convincing claim that the uses of technology will depend on imagination, which thus involves "a moral odor" (*PB* 155), the bomb is grimly real in a way that threatens to overpower Williams's metaphors. We are told that, like love or the poetic imagination, the Curies' and Einstein's scientific creativity resulted in destruction; all are equated as ingenious creations or designs: "the bomb / also / is a flower" (*PB* 165). Henry Sayre has noted that when Williams equates cities and flowers, cities are thereby ameliorated.[75] But in "Asphodel," it is not clear that the bomb can be, or that Williams wants it to be, domesticated in the same way. Indeed, Williams points out that the metaphorical destruction he invokes and even the triumph of the imagination that can discover radium or call the bomb a flower pale next to the actual destruction with which the poem tries to come to terms:

The poem
if it reflects the sea
reflects only
its dance
upon that profound depth
where
it seems to triumph.
The bomb puts an end
to all that.

[*PB* 165][76]

Even more than with his metaphorical identification of poems and machines, Williams's meditation on atom bombs nearly backfires because of the attention commanded by the bomb. Moreover, Williams includes a recognition of this in his meditation. He writes, also in Book II of "Asphodel":

> The mere picture
> of the exploding bomb
> fascinates us
> so that we cannot wait
> to prostrate ourselves
> before it.
>
> [*PB* 165]

The genuine power of a poem such as "Asphodel" lies in its graphic presentation of the need to use language and thought of the sort poetry embodies to come to terms with the world: "I . . . keep on talking / for I dare not stop. / Listen while I talk on / against time" (*PB* 154). In some passages, most notably those concerning the bomb, Williams appears to be talking against time in the sense of trying to turn back the clock or to protest certain aspects of the modern age, perhaps talking specifically against the products of what Lewis had called the physics of time. Elsewhere, however, the poem affirms that it is talking against time in the sense of talking as time runs out. The poem is most immediately about Williams's infidelities and his need for his wife's forgiveness, but Williams also explores the interconnections between the multiple spheres in which he and his readers live—the personal, the political, the biological, and the historical. Not claiming to have effective power in an immediate, practical way over the facts of aging or atomic bombs, "Asphodel" nonetheless successfully celebrates the real value, and the moral odor, which is "no odor / save to the imagination" (*PB* 182), of "fill[ing] up the time" (*PB* 159) with thought and language.

In his later theoretical writing, Williams did not abandon his thoughts on structure or on objects. But he added to his ideas on structure ideas gained from Einstein and Whitehead. "The Poem as a Field of Action" reviews the simple use of technological images in poetry, and sees "the new physics taking its [that is, industrial science's] place" (*SE* 282). Williams's use of industry was never simple, however, and in his mature poetics the new physics may be said to have subsumed both technology and physics rather than replacing one with the other. The new science, especially as taken from Whitehead, proposes that reality, including objects, is "a flow of events";[77] that there is no discrepancy between a celebration of process and an attention to structure or even between science and poetry. Hartley's attempt to seize "the mysticity of the moment" or the interrelatedness of things, which informed *Kora in Hell*, came to be sanctioned by the new science and was shown not to be opposed to the clarity of focus of scientists or to the hard-edged lines of the Precisionists' art.[78]

The mixed style of "Della Primavera Trasportata Al Morale" (in which typographical emblems, menus, and road signs are held together in a stream-of-consciousness narrative that anticipates *Paterson*) seems already to rest on Williams's new vision of how physics dictated a style in which disparate, apparently discrete things coalesce in unique events. Similarly, "The Sea-Elephant" (part of the "Primavera" group) mocks those who would isolate objects, here the sea elephant, or who would distort the animal, which can be known only in its context of enormous flux, the sea. Interestingly, this knowledge of the animal's proper place is voiced by someone practical; theoretical science, poetry, and the American respect for objects finally concur. As a sea creature, of course, the sea elephant cannot be understood; his natural language—"Blouaugh!"—cannot be interpreted. Indeed, the narrative glosses are as suspect as any other voice presented in the poem; more so, perhaps, in that the lines tend to be in overly poetic language, as in the lines "my / flesh is riven" (*CEP* 71). What the creature requires is St. Francis's common stem, some way of being seen or expressed in its natural context without being converted, as Williams puts it (*SE* 29).

The style of the poem provides just such a context. Unlike those who see only one distorted facet of the sea elephant, Williams has his mixed voices and rapid transitions, including several puns, hold the object as "playfully" as the sea (*CEP* 73), identifying the animal seen in its proper, multiple context as love, just as he would say that relativity is "like love" (*SE* 283), a glue by which everything is interrelated.[79]

"The Crimson Cyclamen" (1936), a memorial poem for Charles Demuth who died in 1935, also maintains an insistence upon both mind and object, mental process and structure, multiplicity and design.[80] Like "Young Sycamore," the poem describes the emergence of form "from / subterranean revolutions / and rank odors [moving from roots to stems] . . . dividing, thinning . . . [to eccentrically formed petals, each of which] bends backward" (*CEP* 399–402). Although bound to their origins and bending toward the roots, the petals open to "no completion / no root / . . . color only and a form—" (*CEP* 403). Like "Young Sycamore," too, "The Crimson Cyclamen" describes not only a work of art (here one by Charles Demuth) but also, by analogy, actual plants, life cycles, and artistic creativity.[81]

Yet in "The Crimson Cyclamen," the form of the poem itself is not regular and does not invoke structural descriptions of creative process, although it does celebrate botanical and other structures. Williams begins with seven verses with lines of roughly equal length, but the verses themselves range in size from thirteen to twenty lines each. Williams then breaks the verses into quatrains; four quatrains are followed by three couplets followed by twelve more quatrains, ending with eight longer blocks of verse ranging in length from six to ten lines each. The middle of the poem describes the petals opening in language that recalls "Della Primavera Trasportata Al Morale": "the petals undone / loosen . . . . The flower / flows to release—" (*CEP* 400–01). The poem further equates fluidity and loosening with passion (*CEP* 400 and 403). Appropriately, the passion referred to is multiple and includes erotic passion, Williams's friendship with Demuth, and the creative process. Significantly, the look of the poem on the page also loosens at this point, as Williams abandons the large blocks of print with which he begins the poem for quatrains and, in the lines quoted, couplets, leaving more white space on the page. That is, the visual arrangement of the poem mimetically follows the theme. It is designed, but the pattern is deliberately irregular. Thus, just as "The Sea-Elephant" presents a multiplicity of speakers and images held in a single poem, so too "The Crimson Cyclamen" in form and content insists upon both passion and structure, but it implies that what holds the poem together and what its form mimics is passion.

Although Williams at times described his collages of different voices and images in terms of engineering,[82] his discovery of a new vocabulary for his poetics of process enabled him to resolve some of the tensions involved in describing the value of poems by using technological metaphors. The new physics further allowed him to insist on the importance of poems both as process and as objects. The marriage of object and process is explicitly related to physics in a number of Williams's late essays. A 1948 speech, for instance, alludes several times to Wilson's 1931 chapter on Proust from *Axel's Castle*, in which Wilson writes that "the ultimate units of his [the relativist's] reality are 'events,' each of which is unique."[83] Williams takes from Wilson, and from Whitehead, two further points: first, that there can be "a possible relationship between a style and a natural science" (*SE* 287) and, second, that everything "in the social, economic complex of the world at any time-sector ties in together" (*SE* 283). Thus, science and poetry are *scientifically* related, and Einsteinian physics can underwrite not only Williams's mature style, but also his appropriation of science. There is, of course, a sleight of hand involved in such reasoning. My point is that Williams, and other American modernists, required a sleight of hand in order to write and defend poetry in modern America, and there is at least a poetic justice involved in having Einstein and Whitehead finally sanction Williams's reconciliation, in the 1930s and 1940s, of two seemingly contradictory modernist impulses.

Williams, indeed, almost stopped writing poetry in the period preceding the appearance of the poems I have identified as beginning to reconcile his two earlier styles. In 1924, for instance, he published three new poems; in 1925 no new poems appeared and in 1926 only four poems were printed. In 1932, he explained to Kay Boyle why he had not been writing poetry, telling her that "the form has been lacking" (*SL* 129). Slowly over the next decade, he began writing poetry again, using a style approved by science, yet one that he could claim as having been first the province of poets.

Williams's need to reconcile his two styles may have grown out of more than the necessity that poetry provide a coherent account of itself to a skeptical, modern American world. Malcolm Bradbury's and James McFarlane's study in *Modernism*, for instance, notes the "*bifurcation* of the impulse to be modern," and suggests that the best characterization of the modernist project internationally might be the fusion or "juxtaposing of contradictions for resolution."[84] My point is not that the rapid growth of science and technology in America was the only factor involved in the poets' need for, and difficulty in finding, a coherent defense of poetry, but that in focusing on science and technology, Williams helps reveal an important and often ignored pressure on American modernists, one to which their own choice of images and metaphors calls attention. The crisis that Williams, specifically, underwent from the late 1920s through the 1930s was related most immediately to the Depression; nonetheless, it may be said that the Depression merely underlined the already sensitive issue of the problematic role of the arts in American society.[85]

Williams's attempt to forge a coherent poetics in the particular modern American context to which he, uniquely, paid such close attention reveals both the importance and the difficulty of his project. If some of the questions he asked remain unanswered, his strength lay in his ability to dramatize the difficulties involved in defining poetry's place in America; his struggle itself provides the mirror to modernity in search of which he began his career. Furthermore, the problems Williams raised are still with us, as are the questions his writing keeps alive. He is not only, to quote from Lionel Trilling's discussion of literary history, to "be used as [a] barometer . . . he is also part of the weather."[86]

5

Marianne Moore: The Anaconda-Like Curves of Central Bearings

MOORE'S ideas about poetry are less frequently commented on than Williams's, in part because few of her prose writings are in print and in part because her writing generally is less accessible to readers than Williams's. Yet she was also acutely aware that the place of poetry in modern society needed to be defended. Her editorials for the *Dial* (written between 1925 and 1929), in particular, reveal her developing ideas about how to define poetry's role and function in modern America and her sense that technology had caused the deepest changes in American society.

As Hugh Kenner has pointed out, Moore's "heritage was both literary and technological. Her father's given names were John Milton, and he had suffered a nervous breakdown after the failure of his plans to manufacture a smokeless furnace."[1] Kenner has in mind how Moore's use of the typewriter and of syllabic grids in her poetry relate her art to modern technology.[2] Moore herself, when asked about her view of technology as a source of change in the modern world, said: "It preoccupies me . . . fundamentally and continuously."[3] The images in her poems at times reflect this; for example, she describes a pangolin as "machine-like" (*Comp* 118) and as a "miniature artist engineer" (*Comp* 117).

Moore clearly shared with some of her contemporaries the desire to adopt the aesthetic of modern technology. Thus, she praised a water color by Klee for "resembling a specimen of machine design."[4] Yet she was self-conscious about the claims that art could not thrive in America's technological climate. Her poems appeared in magazines such as *Broom*, which also published Emmy Veronica Sanders's protest against "the manifestation of the American Machine—transforming itself . . . into American Mind; or the mistaking of analytical acumen for *creative* mind."[5] A later issue of *Broom* contained Matthew Josephson's negative description of the spread of machine culture as the Americanization of the world.[6] The small magazines of the period frequently commented on both the lack of creativity and the lack of taste in America, and much was made of, for example, how the United States Customs charged duty on a Brancusi sculpture because they refused to accept it as art.[7]

Moore's notebooks make it clear that she was reading these comments; her poems appeared in many magazines alongside articles criticizing industrial America, and she would certainly have read the exchanges in the *Dial*, where her own work was appearing by 1920.[8] In fact, the first issue of *Broom* includes a description of Moore's poems as technological: "ingeniously constructed, intricate little piece[s] of machinery . . . with cogs and wheels and flashes of iron and steel."[9]

Moore's diaries take issue with a whole series of negative judgments on America. In 1922, objecting to the characterization of Americans as ignorant and provincial, Moore wrote that "we seem a people of *character*, we seem to have energy,"[10] and an unpublished review objected to Harold Stearns's pronouncement on the inferior quality of American arts and letters. Moore concluded of Stearns's scorn for modern American literature: "in as much as Santayana under whose spell Mr. Stearns is—is guilty of similar neglect, one is not surprised."[11]

Moreover, many of Moore's first editorial "Comments" and "Announcements" for the *Dial* implicitly entered the debate on the character and affiliations of the modern American.[12] In a 1925 "Comment" responding to Count Keyserling's *Travel Diary*, for instance, she approvingly cited his statement that "prosperity is regarded as normal in America," and five months later she noted the commonplace association between America and "speed and sport."[13] Moore denied that such an association was unpleasant, distinguishing herself from the critics she was reading. Indeed in 1926, objecting to Karel Čapek's remark that "America's predilection for . . . speed and success, is corrupting the world," Moore appealed to the restlessness of Columbus as the trait responsible for the very discovery of America; she then reminded her readers that women, far from being corrupted by American inventions, benefited from them: "Assisted by the typewriter, the sewing-machine, and the telephone, the American white woman . . . seems as time goes on, more serviceable and less servile."[14]

Moore's years at Carlisle Commercial School, and her first-hand knowledge that women writers, in particular, "need room to experiment and grow . . . and *they need pay*" were clearly part of her unusual acceptance of the trappings of modernity.[15] She shows a similarly atypical acceptance of business. In a 1927 "Comment" on literary neatness, Moore denies that Americans are rapacious, and then adds:

> Pressure of business modifies self-consciousness and genuine matter for exposition seems to aid effectiveness; in for instance, Darwin's scientific descriptions. A similar faithfulness to the scene—to the action and aspect of what makes the scene important . . . —rewards one in the writings of Audubon, the ornithologist.[16]

Moore obviously accepted just those aspects of American life (of business, technology, science, and advertising) regarded with more ambivalence or downright distaste by some of her contemporaries. Not only did she feel such modern enterprises could coexist with literary concerns but she also celebrated the paraphernalia of the twentieth century quite matter of factly.[17] Like Williams, Moore saw machines as tools that left time for writing; unlike Williams, she does not at first seem to have turned her machines into obvious metaphors for American poetry. In short, she seems on first reading unusually careful about separating the practical facts of technological progress from the aesthetic of technology.

On second reading, however, Moore's statements about science, technology, and twentieth-century America begin to look less straightforward. To move, as in Moore's objection to Čapek, from the sonorous association of speed and success, via Columbus, to the usefulness of American technology, which not only serves women but allows them to be *serviceable*, involves a series of imaginative leaps. There is an obvious argument to the effect that technology has provided time for writing and that writing is a kind of service. But what kind of service? And what emotion informs the move from speed and success to instruments of communication and sewing machines? The passage requires that readers pay attention both to what is said and to how it is said, especially to the sound of words. Thus, "serviceable and less servile" echoes, and revises, "speed and success."[18]

Sound was important to Moore. A diary entry notes that "words . . . as well as things, claim the care of an author."[19] In the 1944 "Feeling and Precision" Moore cites Longinus's description of an art " 'bearing . . . the stamp of vehement emotion like a ship before a veering wind,' both as content and as sound; but especially as sound, in the use of which the poet becomes a kind of hypnotist" (*Pred* 8). In her essay, Moore's attention to the sound of words as well as to things suggests that she was not interested simply in praising the practical aspects of American modernity. Rather, she plays with the language ("speed and success") generally used to characterize modern America.

In the process, Moore redefines American creativity as a kind of mental playfulness. Toward the end of her essay on literary neatness, for example, Moore quotes Audubon's assessment of Americans as "rapid"; she then concludes by comparing America with the ostrich. The essay, which begins with technological science and big business, ends with ornithology and speed. The associations here are not unusual. In her later poem on the ostrich, "He 'Digesteth Hard Yron'," Moore says that externalists miss the real and invisible power of the visible (*Comp* 100). In the same vein, Moore's essay celebrates the visible and concrete marks of modern society in order to refer to mental and spiritual characteristics. Her readers are required to think carefully about an apparent untidiness in a discussion of neatness, and to acknowledge that the speed and rapidity alluded to are not merely physical traits.

The idea of speed or motion, to judge from her reading diaries and

Dial editorials, consistently captured Moore's attention and was related on the one hand to an ideal of creative energy and on the other hand to grace and nonpossessiveness. In "He 'Digesteth Harde Yron'," the bird's speed, called heroic, is said to contradict a greed; it does not allow the bird to stand and defend any one place as its own (*Comp* 100). The imagination, too, in moving from object to object, deflects the urge to possess objects materially. Moore's poem, "When I Buy Pictures," talks about not purchasing paintings, but rather regarding one's "self as the imaginary possessor" of objects (*Comp* 48). For Moore, possession by the mind or imagination is not greed. Rather, she celebrates as the real force behind America's success the "pliant dexterity" of the modern imagination, the very motion of which discourages destructive acquisitiveness by concentrating on creativity itself.[20]

Thus, Moore's usual praise of science, technology, and business involves redefinition, and it serves the larger purpose of defending America as a country that might value creativity and imagination. Consistently, her attention to verbal detail reclaims or redefines the technological and scientific virtues of effectiveness, accuracy, and precision.[21] These redefined virtues underlie Moore's invocations of American technology and science. Her comment on literary neatness, for instance, links science and business (the two fields for which she says America is known), and proposes that both Audubon and big business aid effectiveness. But effectiveness, as Moore continues, is like Darwin's or Audubon's "faithfulness to the scene [which, in turn is said to be faithfulness] to the action and aspect of what makes the scene important." Scientific accuracy is then not the accuracy of machines but of observers; moreover, precision and effectiveness include giving value to the world thus viewed; they become forms of imaginative possession, and also activities. Moore wrote that "precision is a thing of the imagination" (*Pred* 4) and "precision . . . creates movement" (*Pred* 141).

Moore's arguments were in part a response to those such as Čapek who identified efficiency and greed as touchstones of a philistine modern America. By her redefinitions, Moore suggested that modern American society fostered, rather than discouraged, creative work. Yet for all her often noted praise and use of science, Moore was sensitive to the further link commonly made between American commerce and American science. One of her relatively rare anti-scientific remarks was made in the context of a discussion about how professionalism and regard for money could be bad for poets.[22] More often, as in her poem "The Student," Moore played with a vocabulary drawn from negative descriptions of American business, and she suggested that profit and advantage, when properly understood, were not to be rejected.[23]

Moore, however, clearly found the accusations leveled at American business worrisome. She felt called upon to respond to Fry's charge that the "world of art also is assailed by a spirit of domination, gainfulness, or expediency,"[24] and she protested sharply that " 'action, business, adventure, discovery,' are not prerogatives exclusively American; and obversely, creative power is not the prerogative of every country other than America."[25] Indeed, the characterization of an exemplary student in "The Student" is taken from a 1931 newspaper account of a man in the arts who resisted the temptations of financial gain. Moore's diary includes a description of Dr. Wilhelm Reinhold Valentiner, director of the Detroit Art Institute and editor of *Art in America*, who is said to have had "the typical reserve of the student. He does not enjoy the active battle of opinion that invariably rages when a decision is announced that can be weighed in great sums of money. He gives his opinion firmly and rests upon that."[26] In short, Moore's celebration of a creative American attitude in "The Student" draws specifically on accounts of Americans who were not exclusively interested in monetary gain; her portrait was a direct response to those who felt American commerce prevented the United States from being a land of culture.

"The Student" may also respond to a 1916 review in the *New Statesman* that proclaimed commercialism above all was responsible for the deplorable faults, including lack of restraint, of American students. "Mr. Leacock Serious" was a review of a book on American education by Stephen Leacock: both T. S. Eliot, who wrote the article, and Leacock blame the dismal quality of American writing on commercialism and the national "hope of the betterment of personal fortune [which] contains in itself an atmosphere in which the flower of literature cannot live."[27]

The first article to appear in print on Moore's poetry noted her "battle against . . . commercialism," but few have pointed out that this position was directly related to the commonplace link between America and commercialism, making Moore's stand part of a continuing debate.[28] Moore's resistance to pessimistic views of American culture is obvious; her exploration of the position of the arts in America, complicated. It seems that her redefinitions avoid the point made by the critics she was reading. Her references to American industrialization, for example, generally shift from images of machine technology to the sciences of biology or zoology, as in her praise of a show by American modernist artists where she begins discussing Strand's machinery as a "perfect combining of discs" and concludes by praising "the anaconda-like curves of central bearings."[29] At the same time, the shift of subject is self-conscious, even humorous. Moore's notes on an early newspaper article about worker compensation being granted for rattlesnake bites repeats the juxtaposition with obvious amusement: "Rattlesnakes infesting a county where road work is being done constitute an industrial hazard."[30] Equating anacondas or rattlesnakes with industry is more than evasion or mere humor. For Moore, the observation of, and human interaction with, rattlesnakes makes them as much a part of the modern, industrialized world as steam rollers or skyscrapers.

Moore's purpose was to emphasize how all human encounters with the world—be they scientific, industrial, or literary—are similar in that they involve reclassifying, revaluing, and so in some sense changing, the world. Thus, Moore's celebrated fondness for naturalists and biologists is still in the service of her defense of American poetry. That is, she argues that America's scientific and industrial feats all rest on the strength of the American imagination. And while the idea that science or technology requires creativity is not original,[31] Moore's close attention to how engineers, businessmen, naturalists, and poets all use imagination in their encounters with the world is unique.

Moore's reading notebooks suggest the sophistication involved in her view of biology or zoology. She noted a quotation from *The Book of a Naturalist* on not praising animals for their virtues or "for their profound knowledge of chemistry [and] . . . higher math as shown in their works."[32] The idea that animal behavior cannot count as virtuous or intelligent seems to contradict the claims made in Moore's poems about pangolins, ostriches, and other animals. Yet a closer look reveals that these poems contain a self-consciousness about the status of their humanized creatures. "He 'Digesteth Harde Yron'" chastises those who make ostriches into symbols of justice at a high cost to the animals themselves. At the same time, however, the poem remakes the ostrich into a symbol of heroism. Moreover, after much sarcasm at the expense of those who capture and kill the bird, the poem captures its ostrich in a remarkably strict poetic form. Moore in fact rewrote the poem, moving from an early draft with eight lines per stanza (in rhymed syllabics) to a poem with seven lines per stanza (still in rhymed syllabics). These revisions—no mean technical feat—indicate that she found the rigors of the form to be an important feature of the poem. The poem's imaginative possession is contrasted with the more malignant ostrich hunters' forms of possession, but still the poem signals the human need to make something of and thus distort or even limit what is perceived. "The Monkeys" makes a similar point about how the poet merely redirects, rather than escapes, the need to make the world one's own and about how this process changes the world.[33]

Thus Moore proposed that poetry can teach the sciences about their common enterprise, namely, inhabiting the world or (what comes to the same thing) making the world inhabitable. Dewey's *Democracy and Education*, which Moore read, opens by stating that it will connect "the growth of democracy with the development of the experimental method in the sciences, evolutionary ideas in the biological sciences, and the industrial reorganization."[34] Moore takes a similar position, but also connects American democracy, as well as the mental prowess exhibited in American science and industry, with the poetic imagination. She in effect repeats the strategy attributed to D. H. Lawrence in her 1921 notes on a review of Lawrence's poems: "Mr. Lawrence does not attack science[,] he puts science into his pocket and walks off with it—a distinct achievement in the direction of unity."[35]

Moore's poems and essays also enact their point as she moves from topic to topic with, to use her terms, ferocity and grace. Her claim is that art "acknowledge[s] the spiritual forces which have made it" (*Comp* 48) by means of its creative transitions, and that such mental agility demonstrates an attractive example of traits such as rapidity, accuracy, and possessiveness, often denigrated as aspects of the American character when practiced in other spheres.[36] Her approval of Kenneth Burke, whom she cited on how the artist "refines the propensities of his age, formulating their aesthetic equivalent," amounts to a comment on her own relationship to her age.[37] The playfulness with which Moore examines and reclaims language is also part of a mental dexterity, an "intellectual wastefulness of aesthetic abundance," that one assumes is, in part, Moore's own aesthetic equivalent to the prosperity of the age, wherein "imaginary possession" means both possession by the imagination and possession that is not real, or not materialistic.[38]

The 1932 version of "The Student" provides an excellent example of the complexity of Moore's acceptance and redefinition of the characteristics of modern America. Here, Moore's complicated appropriation of modernity lies not only in her redefinition of modern American technology, but also in her use of theoretical science, which further underlines her fascination with motion and process.[39] Like Williams, Moore appropriated Einsteinian science as a type of poetic process. She may in fact have drawn on Williams's poem, "St. Francis Einstein of the Daffodils," where Williams associates Einstein with creative process. The issue of *Contact* in which Williams's poem first appeared also included Moore's review of *Kora in Hell*, just following the poem and just before Williams's assertion that "America . . . knows nothing of its debt to the artist."[40]

In "The Student," Moore's American student asks Einstein when "will your experiment be finished," and is told that "science is never finished." Moore, characteristically, is somewhat more straightforward than Williams, yet equally insistent that the processes of science are one indication of American creativity. She replaces Einstein with Audubon, another adoptive American, who "taught / us how to dance . . . how / to turn as the airport wind-sock turns without an error; / like Alligator, Downpour, Dynamite, / and Wotan, gliding round the course in a fast neat / school." Moore's own turn from Einstein to Audubon to aviation and race tracks adds to her point about the accuracy of mental turns the implicit suggestion that the sounds of poetry (*error . . . Alligator, Downpour, Dynamite,* and *Wotan*) can offer a vitality and pleasure in the real, which turns the greed one might otherwise associate with race tracks, which most Americans attend in order to gamble on horses, to a celebration of imaginative gain.[41] In the same way, Moore turns from "speed and success" to women made "more serviceable and less servile," in the early "Comment." In "The Student," Moore portrays America adopting and learning from—but not exploiting—the world's men of science, just as the poem in its final form adopts and footnotes quotations from a lecture, three college mottoes, *The Arabian Nights,* Einstein, Goldsmith, Emerson, Burke, and Henry McBride (*Comp* 278). As Moore writes in "New York," what characterizes us "is not the plunder, / but 'accessibility to experience' " (*Comp* 54).

All versions of "The Student" open with a comparison between French and American education, and insist that in America "one degree is not too much." The later version of the poem suggests that with "us, a / school . . . is / both a tree of knowledge / and of liberty." The earlier version of the poem also equates American liberality, knowledge, and liberty, but less explicitly, and it demands that readers follow the mental turns of the poem and so experience a kind of education in the process of reading.[42]

Moreover, the poem invites consideration of the etymology of words such as education (a leading out, and so a kind of liberty) and science (from *scire, to know,* and so a kind of knowledge). "It is a / thoughtful pupil has two thoughts for [a] word," Moore writes. The poem thus raises questions about the relationship between education, liberty, science, and knowledge. Moore also invokes the commonplace association of democratic America with science and with inferior education.

In "The Student," these issues are related most clearly to Moore's readings about American education. Like Williams in *The Embodiment of Knowledge,* she responded to contemporary debates about educational theory and practice. Specifically, Moore's reading diaries suggest that she found debates about whether or not American education should teach facts or methods to be related to the question of whether science or art was America's forte, and so to the question of whether or not America could produce creative work. Again, Moore seems to have been most attracted to the idea that Americans, contrary to popular belief, valued inventiveness over utility. A quote copied in March, 1916, over which Moore later wrote Randolph Bourne's name, suggests that being a student is related to the creative development of an individual, rather than to mere technical training: "The whole object of education should be to know what one truly and wholeheartedly likes and wants."[43] Moore did not necessarily read Bourne's request in the July 1916 issue of *Atlantic Monthly* for "an investigation of what Americanism may rightly mean," but she asks the same question, along with related questions about the place of art and science in America, in "The Student."[44]

From the same period, Moore was taking notes on Dewey's writings about the necessity of fostering creativity in American schools,[45] and, from Berenson, on how "we value [education] not for its direct results but for its direct effects upon . . . the man who is engaged therein."[46] On the top of the following page in Moore's diary she wrote: "Education . . . the term refers to the spiritual."[47] A slightly later entry notes that students who gather facts only in notebooks act like "fools," since they have "knowledge which can be stolen from them . . . or . . . eaten by rats or by worms or destroyed by fire or water."[48]

Moore's readings about science yielded passages that echo her notes on American education.[49] She quoted a series of statements to the effect that science is an open-ended process: "It is certain that never, before God is seen face to face, shall a man know anything with final certainty."[50] She also described how Duns Scotus answered the question of whether theology is a science by calling it "more properly . . . a *sapientia*, since . . . it is rather a knowledge of principles than a method of conclusions."[51] As Bonnie Costello points out, commenting upon Moore's use of this passage in "To A Snail," Moore differs from her source in identifying science with *sapientia*.[52]

Moore characterized science then not as primarily practical, but rather as endless investigation, indicative of a kind of spiritual energy. Thus, "The Student" lists both theology and biology as sciences. As Moore said later, citing an article by Bronowski from the *Saturday Evening Post*, "science is the process of discovering. In any case it's not established once and for all: it's evolving" (*R* 273). Science and poetry are unified, both being described as processes of discovery. For Moore, science and poetry both also involve discovering a world in flux. That is, Moore's interests in Darwin, Burbank, and etymology are related, in that she equates botanical, zoological, and semantic facts as evolving rather than static.[53]

"The Student," then, takes what C. P. Snow has called two cultures and invites us to see them as related, or twins:

> In each school there is a pair of fruit-trees like that twin
> tree
> in every other school: tree-of-knowledge—
> tree-of-life—each with a label like that of the other
> college:
>
> *lux*, or *lux et veritas, Christo et ecclesiae, sapiet*
> *felici*, and if science confers immortality,
> these apple-trees should be for everyone.

Ironically, the poem is not easily accessible despite its celebration of a knowledge that is "for everyone." But it does resist the labels that would separate intellectual knowledge and science from lived experience: "The football huddle in the / vacant lot / is impersonating calculus and physics and military / books; and is gathering the data for genetics." Like Williams, Moore proposes the embodiment of knowledge. She further resembles Williams in resisting the commonplaces that pitted poetry and intellectual learning against the practicality of science and the American character. In "The Student," science, or knowledge, is equated with vitality and openness, and is thus not incompatible with poetry. A late essay makes the same point, calling Auden one "whose scientific predilections do not make him less than a poet—who says to himself, I must know" (*Pred* 86–87). Moore's diary for the years 1916–1921 also includes, along with her notes about American education, a passage on how odd it is to find "intellectual learning . . . set up in a sort of supernatural opposition to practical wisdom and the results of science."[54]

In 1920, Dewey wrote, "Surely there is no more significant question before the world than this question of the possibility and method of reconciliation of the attitudes of practical science and contemplative esthetic appreciation."[55] While Moore may not have read this particular passage by Dewey, she appears to have attempted just the reconciliation for which he calls.

Yet while "The Student" insists upon the mental agility of American scientists and students (as well as of poets and readers), it also insists upon facts, including the sociological or cultural fact that in America science was seen as offering hard facts. Although she argues that science is primarily creative, Moore at the same time suggests that poetry is practical. In this vein, in a 1935 speech given at Bryn Mawr, in which Moore read "The Hero," originally part of the same poem as "The Student," she compares poetry, music, and math as being of little practical use. Yet the speech, which recommends students read Whitehead's *Introduction to Mathematics*, continues: But "we have no use of applied mathematics that is not based on theoretical mathematics," and concludes that physics, by virtue of improvements to the phonograph, has trained poets' ears.[56] As with Moore's college teachers, one is tempted to accuse Moore of obscurity here.[57] But a look at the context of her speech helps reveal the point, which is that poetry, like math, only appears impractical. We know that mathematics informs life; the implication is that poetry is equally implicated in the real world.

In keeping with this claim that poetry and practicality are not opposed, "The Student" maintains the common definition of *fact*, as indicated in the list of zoological tidbits (" horned owls have one ear that opens up and one / that opens down ") and linguistic nicetics (" Swordfish are different from / gars, if one may speak of gars when the big / gamehunters are using the fastidious singular"). As the poem says, after celebrating mental and aural "going[s] round . . . there is more to learn." So Moore's student is called upon to know facts as well as to be creative. The syllabics in which the poem is written add to the sense that the poem involves an almost mathematical precision. Indeed, the precision with which words are used in the poem is, for Moore, related to science. Her review of Roget's *Thesaurus* pronounces the investigation of words to be "analogous to the laboratory scientist's classification of species in botany or zoology."[58] And, later, she described the form of her poetry in scientific language that almost overturns her usual insistence upon the part played by the human imagination: "I never 'plan' a stanza. Words cluster like chromosomes, determining the procedure" (*R* 263).

Finally, however, Moore's attitude toward facts, in the commonly understood sense of the word, is complicated. Her diary quotes statements that she disagreed with as well as ideas she found attractive. But it is with apparent approval that she cites W. H. Wright on *The Creative Will:* The "artist sacrifices minor scientific truths to his creative inventiveness because he is ever after a profounder truth than that of accuracy of detail."[59] She also copied such passages from W. H. Hudson's *The Study of English Literature* as "the poet does not give what they call facts" and poetry "is an interpretation of life from the point of view and through the medium of the feelings."[60] Moore's emphasis upon the creativity of science and the mind's ability to give value to the world would seem to ally her with Hudson and Wright. In fact, she used the quotation from Hudson in "Picking and Choosing," where she says that Hardy, as poet and as novelist, is "one man / 'interpreting life through the medium of the / emotions.' If he must give an opinion, it is permissible / that the / critic should know what he likes" (*Coll* 52). Echoing also the passage attributed to Bourne in which the object of education is said to be knowing "what one truly . . . wants," Moore suggests that opinions are related to emotion and interpretation; they are a matter of feeling rather than of scientific verification.

Yet "Picking and Choosing" goes on to mock, albeit affectionately, the mistranslation of *summa diligentia* as meaning Caesar crossed the alps "'on the top of a / diligence'. We are not daft about the meaning, but this / familiarity / with wrong meanings puzzles one" (*Coll* 52). "The Student" reiterates that no fact of science "might / not as well be known; one does not care to hold opinions that fright / could dislocate."[61] Moreover, Moore's diary includes a quotation on how poetry that ignores facts "in the long run . . . must have a 'weakening effect on the mind'."[62] And, two pages after she copied passages that argued poetry need not be factual, Moore wrote in her diary the draft of a letter to *Poetry Magazine:* "Dear Poetry, There is a crying need for a Poet's Handbook of Science. W. R. Benét, for instance should be informed that bats do not hang in barns at night, that they fly around at night; . . . Lola Ridge that . . . jaguars do not inhabit deserts."[63]

Science may be viewed as creative, but poetry takes on the practical commitment to hard facts that Moore's redefinition no longer identifies with science or education. "The Student" remains an attempt to unify two cultures in the process of producing a defense of American creativity. But Moore's usual and more traditional humanistic poetics—implied in her celebration of creativity and process—rest uneasily with her occasional attempts to appropriate the hardness, even the impersonality, of American science and technology for poetry. Moore seems to have sensed this problem; her work refuses to settle for easy answers. The early version of "The Student" is best characterized by its continual refusal to rest with any particular redefinition of science or poetry, although it insists upon the process of redefinition. The poem keeps open the distinctions between knowledge and life, fact and creativity, poetry and science, even while insisting that they cannot be neatly separated.

Absolute knowledge, presumably, precludes the necessity of continued investigation. Thus "The Student" maintains its valorization of process and imagination in refusing to settle the questions it raises. On various levels, the poem courts the uncertainty that sparks the imagination and intellect. It suggests we can neither separate nor wholly merge the terms of Moore's dialectic; we can neither give up trying to get things right, nor will we ever finally get things right. The language itself repeats this suspension. The student interested in a stranger's résumé is pleased to be told that "science is never finished," but the poem has just insisted upon the etymology and context of words, specifically French words—*valet*, *bachelor*, *damsel*—reminding a thoughtful reader that résumé means a "summary of experience, or conclusion," yet also comes from the Latin, *resumere*, meaning "to take up again" or "to continue."

The dialectic or suspension set up in "The Student," which recalls Dewey's and Williams's understanding of the process of thinking, is characteristic of Moore.[64] She announces in a single poem that science is not inert fact; poetry, science, and education in America are all concerned with creative energy and process; but facts are still facts and not to be ignored. Similarly, as a passage in Moore's diary put it: "Education . . . the term refers to the spiritual or bodily effect of a course of experience be its nature what it may . . . [In viewing art, it] means that the mind rests in its object . . . beholding it without deserting it."[65] The mind's relationship to the object of its attention is dynamic and creative, but does not entail deserting the object.

In her article announcing that Williams had won the *Dial*'s award for 1926, Moore wrote: "Williams is a doctor. Physicians are not so often poets as poets are physicians, but may we not assert confidently that oppositions of science are not oppositions to poetry but oppositions to falseness?" (*Pred* 135). For Moore, both science and poetry, variously defined, claim truth as an uneasy marriage of continued endeavor and fidelity to the world, allowing us neither to ignore facts nor to ignore the effects of human observation and evaluation. This double focus on object and observer is repeated in "When I Buy Pictures," where Moore insists that art must both reveal "the spiritual forces which have made it" and give " 'piercing glances into the life of things' " (*Comp* 48).

As in Williams's poetics, such a commitment to both mind and world, process and fact, could be problematic.[66] Moore did not reconcile these poles, and apparently felt an ethical commitment to both sides of the dialectic. That the poems and essays are not easily accessible has less to do with the lack of relevance of which she is sometimes accused than with her dual attention to mind and matter, with her tactic of redefinition, and, one might add, with a lack of condescension. Moore tells us, again in "The Student," that "we are / as a nation perhaps, undergraduates." Her poems exemplify rather than preach certain modes of understanding, and thus they school us. In her later version of "The Student," Moore describes a student who is "a variety / of hero": "he renders service when there / is no reward" (*Comp* 102). Moore's ideals, like her heroes, were various, but she consistently argued for one service she would have poetry render, namely, the encouraging of respect for what is (that is, for concrete particularity) without relinquishing the need morally and imaginatively to redefine values, in a constant engagement with the world that is the true moral center of her work.

Moore's difficult vision of science's and poetry's roles in America returns us to the larger question with which "The Student" is concerned. If Moore's defense of poetry is in terms of science and fact, I have argued that this is in part because criticisms of American society had raised the problem of how to launch a defense of American poetry without ignoring the commercial and industrial values for which America was famous. To cite Lewis Mumford's essay on the city in Stearns's 1922 *Civilization in the United States*, the "highest achievements of our material civilization . . . count as so many symptoms of its spiritual failure."[67] To counteract such views, Moore pointed out the spiritual in the material and the complex interrelations of higher learning and life, fact and imagination, poetry and science.

Yet Moore did not pretend that American society's potential—great as she claimed it to be—was all necessarily for the good. Despite my argument to the contrary, there is a grain of truth to Bernard Engel's judgment that "The Student" is about American immaturity and to Randall Jarrell's remark about Moore's relationship to modern America: "she accepts her own society scarcely more than Cato accepted his."[68] The poem ends:

> . . . in this country we've no
> cause to boast; we are
> as a nation perhaps, undergraduates not students.
> But anyone who studies will advance.
> Are we to grow up or not? They are not all college boys
> in France.

The modulation of tone in these final lines is complicated. Not boasting—having humility—is a trait Moore admired, but having no cause to boast can mean either that Americans are not bound to some single cause, that is, they are open-minded, or it can mean that a celebration of America is premature.[69] In the same vein, the French clearly are not all college boys because some have grown up and are (perhaps in contrast to us) students in the best sense of the word.[70] Yet the comment cuts two ways. We are referred back to the beginning of the poem and reminded that the French system of education is not democratic and so disenfranchises many. By contrast, American education may be better.

The poem itself is open-ended in that it refuses to settle such issues. American democratic education may be (like the poem) open-minded, broad-based, even humble. Not to be a graduate is not necessarily to be uneducated; it could indicate that education never ends, being a matter of degrees in the sense of increments rather than of sheepskins. Yet it is also true that to be undergraduates may indicate immaturity and a lack of self-consciousness, neither of which would count as a virtue.[71] A number of the virtues of Americans presented in the poem are highly qualified. For example, "impersonating calculus," as the poem's football players do, may be to give life or body to abstract disciplines, but it may also qualify the genuineness of such a calculus. The poem continues: "If scholarship would profit by it, sixteen / foot men should be grown; it's for the football men to / say." While Burbank's genetic experiments intrigued her, Moore's comment on football men is surely tongue-in-cheek, especially in light of her self-consciousness about the relatively inhuman, and unvaluable, status of the unobserved world. Some observers might be able to see scientific theories at work in a football game, but like the animals who cannot be praised for their knowledge of chemistry, the football men are certainly not best qualified to say in what scholarly profit might consist.

Moore was aware that many of her readers would understand advancement in material terms, rather than the intellectual and spiritual advancements obliquely proposed in the poem. Yet, by forcing us, as readers, to observe the science put into practice by football men, the poem tries to educate its audience. If the poem's football players are not the American students for whom the poem calls, those who have read Moore's poetry are being prepared to fulfill the potential that "The Student" outlines as distinctly American.

T. S. Eliot's characterization is apt: "Moore's relation to the soil is not a simple one."[72] We may build on Eliot's insight, taking the soil to indicate the things of the world (facts, animals, and so on) or the specifically American climate in which Moore wrote. In her redefinition of science, fact, accuracy, speed, and the vocabulary of profit, she revalues the common characterization of technological and industrial America, making this revised portrait of the country the foundation of her vision of modern American poetry. More self-consciously than many of her contemporaries, Moore both claimed and resisted the traits she and others found in modern America. To set imaginative agility and possession against more destructive manifestations of the same impulses—or, rather, to redefine the field in which such impulses should operate—is an attractive political stance. Moreover,

Moore's clarity about her poetic project allowed her to avoid the confusion that Williams, for example, at least in his earliest writing, exhibited in his simultaneous attraction to and mistrust of his age.

Yet this strategy does not solve the dilemma American culture presented to its poets in the early twentieth century, as witnessed by Moore's self-consciousness about how to reconcile American practicality with creativity, or accuracy with imagination. At least as a method of reforming her society, which is a primary goal of her redefinitions of American values and of her insistence on the relationship between public and poetic language, the effectiveness of Moore's strategy is open to question. Her ultimate insistence on the primacy of the individual mind's confrontations with its world may have prevented her from going on to offer analyses of such questions as why women as a class might be served by technological advances or why certain character traits are fostered by postindustrial societies. Moreover, the valorization of speed and efficiency as imaginative virtues may have masked the need to question such values in other spheres; one must recall that as Moore wrote, efficiency became the battle cry of American business, inspired by Taylorism, and so was a bone of contention between labor and management.[73]

Williams more obviously paid attention to labor complaints, as witnessed in his writings from "The Strike" section of "The Wanderer" through his short stories of the 1930s through *Paterson.* His inclusion of Allen Ginsberg's implicit accusation in *Paterson*—"Do you know this part [the political life and the bars] of Paterson?. . . I wonder if you have seen River Street" (*Pat* 194)—self-consciously asks how the America of his poetry could be mapped onto the harsher realities of industrial American life.[74] Yet for Williams, as for Moore with her concentration on clearly seeing or morally redefining American traits, the poetic strategy used may have distracted attention from the need for more active social (rather than individual) change in order to achieve a society that would embrace the human values both poets associated with the arts.

Further, it is not clear that Moore convinced even herself that her partial strategy for the defense of poetry would open a place for modernist poetry in America. She maintained her ambiguous identification of poetry with hard facts, as well as her denial that science should be valued for its hardness. Moreover, she felt the continued need to argue with those who scoffed at modernist experiments.[75] Finally, her view of Williams's and Stevens's poetry as proof that "poetry in America has not died" (*Pred* 139) suggests that she felt the proposition needed proof.

Such problems, however, affect the modernist project as a whole; Moore's critique of modernity as well as her clear-sighted, and subversive, redefinition of American modernity still stands as one of the most intelligent poetic responses of her time. As with Williams, the dilemmas Moore confronted fueled her poetry, resulting not only in an astute critique of modern America, but also in a poetry, the quality of which may be the best defense of Moore's claim that American modernity could foster American poetry.

Moore's self-consciousness about how to maintain intellectual honesty and imaginative rigor, while still writing a poetry democratic not only in what it included but also in being accessible to an American audience, may have proved less beneficial to her writing.[76] Still, the reasons for Moore's later turn to a less difficult—and most would agree lesser—poetry, presented in more popular forums, remain to be investigated more fully. The fact is that Moore's attempted reevaluation and defense of American modernity, including its business, technology, and science, is what yielded some of the most compelling poems of early twentieth-century America.

6

Wallace Stevens: Getting the World Right

WILLIAMS's desire to give voice to the American experience led him at times to associate his poems with other American products, at times to say that poets were in some respects like engineers or scientists, and at times to claim a scientific foundation for his poetics. Moore also insisted that poetry was comparable to more respected professions, but she was perhaps more self-conscious about her attempt to redefine those fields to which she compared poetry. When asked if she considered herself an American poet, Moore responded affirmatively and then explained she thought of herself as "an American chameleon on an American leaf."[1] She never simply took on the protective coloring of her environment, however. Rather, she attempted to change that environment in the process of defining a public role for poetry in America.

From the first, Stevens criticism has raised the question of Stevens's relationship to the American context in which he wrote.[2] As he himself said in response to the same questionnaire that called forth Moore's image of herself as a chameleon, his poetry is not "flagrantly American."[3] Recently, however, a number of critics have turned to the archives at the Huntington Library and elsewhere to reveal that Stevens's writings are more deeply rooted in an American context than they first appear.[4] These studies help show that, although Stevens was less outspoken than Williams or Moore about his interest in American culture, he too was strongly influenced by the need to relate poetry to the larger social and political world of America. Like Williams's and Moore's, Stevens's work was in part shaped by his reaction to American attitudes toward poetry and science. Stevens, like Williams, responded most strongly to the particular difficulties of defending the importance of poetry during the Depression years, and ultimately found that the new physics offered a way of envisioning the poetry he wrote as grounded in the *real* world.

Stevens may be the American poet who made the most serious use of the new physics, especially of quantum theory, although he paid less attention to technology or to the commercialism associated with American industrialization than many of his contemporaries. Only rarely did Stevens participate in the widespread identification of American modernism with the aesthetic of technology. Still, he was not unaffected by the ideas with which he came in contact through his friendships with Moore, Williams, Kreymborg, and members of the Arensberg circle, or through his reading of the *New Republic*, *Poetry*, *Hound & Horn*, *Broom*, and the *Little Review*.[5] Indeed, between 1914 and 1923, he was apparently attracted to a machine aesthetic. Glen MacLeod suggests that some of Stevens's early, gem-like, poems were responses to Duchamp's readymades, and that lines such as "Clasp me, / Delicatest machine," from "Romance for a Demoiselle Lying in the Grass" (*OP* 23), owe a debt to Duchamp's *The Bride Stripped Bare By Her Bachelors, Even*.[6]

Stevens's letters and journals also include some positive descriptions of the beauty of the urban environment that he more often ignored or disliked. In 1909, he described the "Brooklyn Bridge . . . brilliant with its hundreds of lamps. And the Singer building with its lighted tower [shining] far more beautifully . . . than you would expect."[7] In a letter written later the same year, Stevens noted that a new Manhattan bridge was "a mass of steel, suspended over the East River by steel cables attached to two lofty steel towers. A chant to the builders! . . . [the city] is superb. It may not be beautiful, but in force and strength it is superb[;] yet I think it *is* beautiful. Its power is inspiring."[8] An earlier journal entry describes the stars as geometrical, and adds: "I rather like that idea of geometrical—it's so confoundedly new!" (*L* 48). Although many other things also inform the poem, the 1922 "Stars at Tallapoosa" (*CP* 71) returns to this idea.

In spite of these indications that he could share the modernist fascination for technology and the accompanying praise for the originality of those who created it, Stevens was not usually taken with the look of modern technology or of the modern urban landscape. He lived in New York from 1900 to 1916, and aside from a few notes calling New York "the greatest place to be Americanized" (*SP* 74) or admitting a grudging admiration for the "electric town" (*L* 52) in which he lived, his journals and letters are remarkable for how little they say about what was then one of America's more modern cities.[9] Stevens wrote much more about the weather, birds, and flowers he encountered during his long walks into the New Jersey countryside than about the city itself.[10] In passing, he reveals, if anything, his distaste for the crowds (*L* 141) and the commercialism (*L* 38 and 42) of New York.

In general, then, Stevens did not share the excitement that many of his contemporaries voiced about the look of urban modernity, which he described negatively in an early journal entry as "Chicagoan" (*L* 32). Even his rare admiration for buildings and bridges was often qualified: praising the Washington Bridge involved describing it as Roman (*L* 94), that is, as reminiscent of the glories of the past rather than as distinctly modern. In fact, far from appropriating modern technology like so many of his peers, Stevens expressed his mistrust of it and of those who practiced "professional modernism" (*L* 647). At no point in his career did Stevens use a sustained metaphorical appeal to technology like the one found in Williams's suggestion that a poem is "a structure of little blocks," an idea Stevens disliked (*L* 803).

Yet the idea that poetry needed to be made new did preoccupy Stevens. Like Moore and Williams, he wanted to provide American poetry with a new rationale. In Stevens's case, family pressures forced him to confront what I have argued were common American attitudes toward poetry, particularly genteel poetry. This pressure is evidenced in the Stevenses' objection to their son's desire for a "literary life" (*L* 52–53); an 1899 letter from his father ends: "I am convinced from the Poetry (?) [*sic*] you write your Mother that the afflatus is not serious—

and does not interfere with some real hard work" (*L* 23). A few months later, Stevens's journal contains a passage, in part a response to his father, that presages his lifelong attempt to define the importance of poetry: "Those who say poetry is now the peculiar province of women say so because ideas about poetry are effeminate. . . . Poetry itself is unchanged" (*L* 26).[11] As late as 1913, Stevens still betrayed his partial internalization of the attitude he opposed, calling his habit of writing "lady-like" (*SP* 257). And throughout his life he was unhappy that the "writer is never recognized as one of the masters of our lives . . . [that people feel the] writer is a fribble" (*L* 510). In part, then, he required a defense of poetry in order to declare himself a poet in America. For Stevens, an adequate defense of poetry had to indicate poetry's worldly function or at least some relationship between poetry and the world.

Stevens's new ideas about poetry formed slowly, over a number of years, reaching their fullest expression in a series of talks and lectures written in the 1940s, but evident also in the poetry and letters written in the 1930s. From the beginning, like Moore and Williams, he was unwilling to defend art purely for art's sake (*L* 24).[12] In his earliest writings, Stevens cast about for a defense of poetry. An 1899 journal entry shows him shedding attitudes characteristic of the nineties: "Art . . . detached, sensuous for the sake of sensuousness, . . . is . . . rubbish" (*L* 24). Even as he dismissed art for art's sake, Stevens continued to insist that "the real use of . . . beauty . . . is that it is a service, a food" (*L* 24).[13] As if afraid that beauty for its own sake might be perilously close to the position he was rejecting, he concluded: "Art must fit with other things; it must be part of the system of the world" (*L* 24). Stevens, then, continued throughout his career to question exactly how poetry fit with other things.

Perhaps because of his apparent lack of interest in identifying himself with the aesthetic of technology, or with American urban modernity, Stevens was unusually resistant to the idea that poetry's defense should be in terms of utility. Charles Mauron's 1935 *Aesthetics and Psychology*, a book that Stevens carefully read and annotated, apparently reinforced this idea.[14] Mauron suggests that scientists are intent upon future action, while artists have as an end self expression and the creation of a language: "The artist . . . has . . . no interest in the practical efficiency of the analogies he discovers."[15] Stevens underlined the last phrase, and his marginalia repeated the emphasis on the uselessness of art.[16] Indeed, Marianne Moore recognized Stevens's unique "pride in unservicableness [*sic*]," perhaps thinking of a 1934 interview in which they both participated where Stevens said he did not like the word "useful."[17] Earlier, too, he distanced himself from the "modern conception of poetry [which] is that it should be in the service of something" (*L* 147).

If Stevens distinguished himself in consistently disclaiming utility as a measure of poetry's value, the steadfastness with which he voiced these disclaimers acknowledges that he felt poetry needed to be defended against fields, such as technology, in which practicality was a gauge of value. Moreover, he did not abandon the insistence that poetry had real value. This seems to have been his point in answering the second question in the 1934 *New Verse* interview, namely, whether there could now be a use for narrative poetry. Instead of balking at the question, Stevens responded that there "can now be a use for poetry of any sort" (p. 15).

Stevens often called for writing that would have an allegiance to the earthly, to what he called life or reality. An early journal entry noted that sonnets "have their place . . . but they can also be found tremendously out of place: in real life where things are quick, unaccountable, responsive" (*L* 42).[18] Similarly, Stevens approvingly commented on one of his own journal entries, claiming it exemplified the "quick, unexpected, commonplace, specific things that poets and other observers jot down in their note-books. It was certainly a monstrous pleasure to be able to be specific" (*L* 29).

Thus Stevens, like Moore and others, calls poets observers of the commonplace world. If Moore's paradigm was biology or ornithology, Stevens's model at first was botany. His journal for 1899 notes that the large abstractions that fill one's mind upon visiting the country—"Freedom, beauty, sense of power, etc."—quickly pale and are "in turn succeeded by the true and lasting source of country pleasure: the growth of small, specific observation" (*L* 30). And Stevens's journal frequently contains notes that could have been written by a student of botany: "Larkspur is various and is to be known by the rabbithead-like corolla . . . the calix—generally purple, or mixed purple and pink" (*SP* 44). The reality Stevens wanted to claim for poetry included not only country flora and fauna but also the hustle of twentieth-century living. A 1909 letter about automobiles on the New York streets speaks of going home to read Coleridge, and finding it "heavy work, reading things like that, that have so little in them that one feels to be contemporary, living" (*L* 121).

In spite of the pleasure he took in attending to the world around him, Stevens nonetheless often found the contemporary world, whether revealed in the country or encountered in the city, unpoetic. Although he sometimes admired those who accepted "things as they are," and dismissed idealism as an "intolerant form of sentimentalism" (*SP* 209), he, like Moore and Williams, wavered in his insistence that modern poetry should concern the commonplace and involve close observation of the physical world. At times, he proposed an airier model for poetry, which he aligned with the ideal. One of his early pieces in *The Harvard Advocate* (1899) portrays a daydreaming poet who dismisses "science, . . . economic calculations and mathematical designs . . . [to become] less Faust than Pan" (*SP* 27). A 1902 journal entry proclaimed that he was not a materialist (*L* 60) and later Stevens even wrote to Elsie Moll that he found facts bothersome: "Facts are like flies in a room. They buzz and buzz and bother" (*L* 94). He also quoted Paul E. More on those such as botanists who examine "Nature with microscopes": "I sometimes think a little ignorance is wholesome in our communion with Nature" (*L* 133). The idea of not observing nature too closely is given body marvelously in one letter to Elsie Moll where Stevens dropped his usual sentimental pose to complain that the rose she had sent him had "little yellow things in it— . . . creepy" (*L* 146). He also wrote, "If only it were possible to escape from . . . Facts" (*L* 150); in other words, in spite of his desire to observe the details of the world, Stevens was also capable of wishing he could escape quotidian reality.

It is often noted that Stevens's poems explore the relationship between imagination and reality, and that they refuse to settle the terms of that relationship. As a 1931 letter states, for Stevens one "of the essentials of poetry is ambiguity."[19] Less frequently noted is the large debt Stevens's vision of reality and his initial puzzlement over what poetry and the imagination have to do with reality owe to his view of science.[20] Along with the *Advocate* piece, Stevens's first notes to himself on the subject of poetry reveal his sensitivity to the antagonism between poetry and fact, and to the idea that scientists might be better observers than poets and have more direct links with the world. His diary recalls a walk taken with two friends, Shearer and Mengel, on July 31, 1899: "taken up with conversations about gale-bugs or gale-flies, ichneumon bugs, . . . etc. You felt in the two men an entire lack of poetic life, yet there was an air of strict science, an attentiveness to their surroundings which was a relief from my usual milk and honey" (*SP* 49).

As Stevens told Hi Simons, when he was young he believed there was "a law of contrasts" (*L* 368), and in contrasting science with poetry he generally placed the ideal nature of poetry against the practical and factual nature of science. That is, in spite of his periodic celebrations of specificity and detail, what Stevens prescribed for poetry often was neither factual nor particular. His obviously long-standing argument with Shearer (one of the scientists on the July 31, 1899, walk) yielded an earlier diary entry: "Shearer may be right about the infinity of facts—but how many facts are significant and how much of the ideal is insignificant?" (*SP* 46). Again, in August of the same year, he wrote: "I believe . . . in the efficacy and necessity of fact meeting fact—with a background of the ideal . . . I'm completely satisfied that behind every physical fact there is a divine force. Don't, therefore, look *at* facts, but *through* them" (*L* 32).

Stevens's contradictory stances were both related to commonplaces found in late nineteenth- and early twentieth-century comparisons between poetry and science. It is not unusual to find claims about poets as observers of the real or of the particulars overlooked by scientific classifications and generalizations. Williams's statement that art, as opposed to science or philosophy, "alone remains always concrete, objective" (*EK* 56) comes to mind. At the same time, one equally often finds poetry associated with the ideal—with Stevens's "milk and honey"—as opposed to science, which is linked with what Whitehead calls the "irreducible and stubborn facts."[21] In all his writing before 1915, Stevens shared the American view of science, whether defined as factual or as the quest for intellectual knowledge, as antithetical to poetry. Yet although he consistently contrasted science with poetry, he insisted throughout that poetry was still in some way part of the everyday, *real* world.

Indeed, before 1915, Stevens's writings contain these contradictory stances more or less unselfconsciously. Poetry would eschew the mere work-a-day scientific facts and celebrate the ideal, the higher relationships between things. Alternately, poetry would involve specific, living reality, and celebrate a Pan-like sensuality in contrast to the dry abstractions of science. By 1915, however, Stevens found a way to fuse these contradictions into a more coherent view that allowed him self-consciously to insist both that poetry concerned the "quick, unexpected [and] commonplace" and that poetry looked for larger interrelationships or *systems*.

This view is one voiced in Mauron's *Aesthetics and Psychology*. The human need to admit both the sensual world and the abstract systems which give the world value is part of Mauron's point when he suggests that science reduces difference and that in science "human reason always tends to bring out the maximum of *identity* between phenomena [a tendency which] . . . comes up against the [more specific] results of sensibility."[22] Between 1935 and 1944, when Stevens read Mauron's book, his ideas about the complex nature of poetry and the world led him to underline the conclusion of the above line of reasoning, namely, that "any conception of the world appears as a compromise between these two tendencies [identity or logical generalizations, on the one hand, and the quirkier results of sensibility on the other]" (p. 91). By 1946, Stevens had copied out for himself, longhand, H. D. Lewis's "On Poetic Truth," which begins with the statement: "Poetry has to do with reality in its most individual aspect. An isolated fact, cut loose from the universe, has no significance for the poet" (*OP* 235).[23] Stevens follows Lewis and identifies individual reality with facts given a context and value by an imaginative observer. The idea is similar to Moore's account of observation. In fact, a 1948 essay by Stevens specifically links Moore and Lewis. Stevens says Moore uses facts to create a *significant* reality, reality given value, in a way that makes her poetry (to quote Stevens) not "factual [but rather] an abstraction," which Stevens equates with individual reality as opposed to isolated fact (*NA* 94–95). Thus Stevens evolves a more sophisticated version of the idea that poetry will pay attention to detail and also to the interrelatedness of things. Stevens's emphasis upon the complex relationship between the particular and the abstract first grew out of his need to confront the contradictory ideas he found himself drawn toward as a student, however. That is, his earliest mature poetry, written before his reading of Mauron, exemplifies a dialectic that only later is defended by appeal to a well-articulated theory.

Nonetheless, Stevens's first published poems are concerned with poetry's relationship to the world. He wrote to R. L. Latimer in 1935: "Everything is complicated; if that were not so, life and poetry and everything else would be a bore" (*L* 303). His early poems can already be seen as celebrations of complexity.[24] In "To the One of Fictive Music," for example, Stevens characterizes the imagination as the consciousness that separates us from the world, and as that which involves the human drive for a clarity and certainty, the urge for an order not usually found in nature.[25] Yet he also suggests that imagination gives us the natural world; in "To the One of Fictive Music," the muse is connected with the earth-as-mother, crowned only with "simple hair" (*CP* 87).[26] The poem describes men "musing the obscure," and the pun is instructive. By the 1920s, Stevens already, like Moore, proposed that we can only understand the world by an act of imagination and yet the very act of humanizing or structuring an ambiguous world is a departure from reality. Even the early poems that call for a stark confrontation with the real acknowledge that "the absence of the imagination had / Itself to be imagined" (*CP* 503). In sum, like both Williams and Moore, Stevens emphasizes the interdependence of reality and imagination from the beginning of his adult career.

Moreover, his view of the flux of both inner and outer reality underlies his view that any specific image quickly becomes passé, like the muse who becomes one "of the sisterhood of the living dead" (*CP* 87). He wrote to Ben Belitt, "Life cancels poetry with such rapidity that it keeps one rather breathless" (*L* 314) and to Harriet Monroe that he found it difficult to "pick a crisp salad from the garbage of the past" (*L* 232). Stevens's aim was to keep the cycle of imagination, which moves to construct, reject, and reconstruct orders, in motion in his language: "There must be no cessation / Of motion, or of the noise of motion" (*CP* 60). This motion provides a response to the quick, unexpected nature of reality already noted in the early journals. Further, the insistence on motion must have been reinforced by theories with which Stevens came in contact through his peers in the world of art and poetry; the celebration of process is voiced not only by writers such as Moore, but even more explicitly by the painters and poets of the Arensberg circle whom Stevens saw frequently between 1914 and 1923.

Stevens's remarks as well as the style he adopted suggest that his early poems continued his insistence that poetry was part of the world. The reality he claims for poetry is twofold. There is, first, the process by which the world is known, including both imaginative projection and the human urge for truths and closure—the "blessed rage for order" (*CP* 130). There is also an implicit appeal to the flux that characterizes the self and the natural world. Life is rapid, as Stevens says; he adds that "the self consists of endless images" (*L* 670). Insofar as human observation and that which is observed are both characterized as processes, Stevens's early identification of poets as observers is not brought into question by the poetics informing his first two volumes of poetry. His early celebration of commonplace, specific details, however, is difficult to reconcile with the poetry and poetics of 1915–1936.

The early poems, of course, do maintain that poetry celebrates the imagination, which in turn gives us our sense of connection with the physical world; the woman in "Sunday Morning" is told that the earth—the "bough of summer and the winter branch" (*CP* 67)—must replace the gods, just as seven years later Stevens told "A High-Toned Old Christian Woman":

> We agree in principle. That's clear. But take
> The opposing law and make a peristyle,
> And from the peristyle project a masque
> Beyond the planets. Thus, our bawdiness,
> Unpurged by epitaph, indulged at last,
> Is equally converted into palms,
> Squiggling like saxophones.
>
> [*CP* 59]

Earthy or imaginative desire, in constant motion, without epitaphs, becomes one of Stevens's articles of faith. The early poems, however, never confuse the reality that they claim imagination can reveal with naturalism.

Martha Strom suggests that "To the One of Fictive Music" specifically enacts Stevens's argument with the localists, that is with the

Seven Arts school as represented in Paul Rosenfelds's *Dial* articles and to a lesser degree with Williams.[27] Her suggestion is reinforced by the two poems Stevens selected to preface his celebration of the one of fictive music. "Gubbinal" begins with a voice that pretends to accept realism: "Have it your way. / The world is ugly, / And the people are sad" (*CP* 85). "Two Figures in Dense Violet Night" begins, presumably with a woman (and muse figure) proclaiming: "I had as lief be embraced by the porter at the hotel / As to get no more from the moonlight / Than your moist hand" (*CP* 85), and goes on to reject the reality of hotel porters and clammy hands. Again, as with the infested roses encountered during his courtship, Stevens recommends poets not look too closely at isolated facts, even though they celebrate the same faculty by which all people live in the world.

Stevens's poems "Botanist on Alp (No. 1)" and "Botanist on Alp (No. 2)" hark back to his earlier arguments with Shearer and describe a poetic replacement for botany that eschews isolated facts for a broader vision, as with the pensive man in "Connoisseur of Chaos" who can see the possibility of a perspective from which the multiplicity of the alps would be "a single nest" (*CP* 216). Stevens's botanists long for a "central composition" (*CP* 135) and bespeak the need for a unifying secular belief to replace religious belief: "For who could tolerate the earth / Without that [celestial] poem, or without / An earthier one" (*CP* 136)? The second poem ends, however, by suggesting that the loss of a single perspective releases the imagination and the tongue, allowing a joy in the creations of language and in projecting oneself into the world; the poem continues proposing an earthier poem: "tum, tum-ti-tum, / As of those crosses, glittering, / And merely of their glittering, / A mirror of a mere delight" (*CP* 136). Although the poems lack realistic detail, they call for a unifying vision that will not deny but rather celebrate earthly multiplicity.

Stevens's decision to retain remarkably little description of the quotidian in the poems published in *Harmonium* or *Ideas of Order* may also have been influenced by his brief career as a newspaper man.[28] The naturalism of the American novel in this period, for example, has been said to owe a debt to the definition of reality fostered by the press in the early part of the century.[29] But as those working in journalism sometimes found, the reality reported by the papers was increasingly a commodity, sometimes literally manufactured by the press.[30] Not only did this underline the problematic nature of a style that claimed to present objective truth, but it also emphasized how news items, like other isolated facts, lacked enduring value, and were both quickly consumed and perishable.[31] In any case, Stevens's encounter with the New York press must have reinforced his mistrust of naturalistic detail as an adequate representation of reality and also increased his resistance to the more programmatic calls for the use of local details in American art and poetry; as he wrote in "Adagia,": "Realism is a corruption of reality" (*OP* 166).[32]

The poetics implicit in Stevens's early poems do entail a theory about why poetry might replace religion as the stronghold of values and why the imaginative process celebrated in poetry was important for the modern world. But he nonetheless worried about how the reality he wanted to claim for poetry was to be equated with the facts of everyday, contemporary life. If earlier nineteenth-century views of science and poetry helped to present him with the problems he considered, Stevens would draw on modern physics to sanction some of his later conclusions about human knowledge, nature, and their interrelatedness. The immediate impetus for rethinking his poetics came, however, from critical responses to his poetry in the mid-1930s.

By then, American critics, having accepted the modern idea that poetry should be part of the real world, faulted Stevens's first two volumes for their distance from the harsh reality of the Depression. Stevens's "mere delight" came to look more trivial than pure (to use two variant meanings of the word *mere*). Theodore Roethke's review of *Ideas of Order*, for instance, found it "a pity" that Stevens did not project his imagination "more vigorously upon the present-day world."[33]

Stevens obviously felt that his critics may have had a point about the unworldliness of his poems. By 1935 he would say of his poetic realist or "inquisitorial botanist" (*CP* 28) from "The Comedian As The Letter C": "It is hard for me to say what would have happened to Crispin in contact with . . . the present-day unemployed. I think it would have been a catastrophe for him" (*L* 295). In 1940, he wrote: "At a time of severely practical requirements, the world of the imagination looks like something distorted" (*L* 372). Still, in the 1930s, he again sharply rejected naturalism, this time as it was advocated in leftist journals such as *The New Masses*.[34] His response to the article by Stanley Burnshaw is probably best known.

Burnshaw called *Harmonium* "the kind of verse that people concerned with the murderous world collapse can hardly swallow today except in tiny doses," and asked, referring to Stevens's second volume, whether the poet would "sweep his contradictory notions into a valid Idea of Order?"[35] Stevens's answer was to include a portrait of Mr. Burnshaw in "Owl's Clover," a poem that contains several unsympathetic characters who call for Marxist art. For instance, Stevens pictures a statue carved with the inscription,

> "*The Mass*
> *Appoints These Marbles Of Itself To Be*
> *Itself*." No more than that, no subterfuge,
> No memorable muffing, bare and blunt.
>
> [*OP* 48]

Such a vision of art obviously ignores Stevens's more elusive imagination.[36] Following the ideas implicit in *Harmonium* and *Ideas of Order*, Stevens reasserts that, given the static nature of literary and political conventions and the flux of the world, naturalism is not the detailed reporting of actual fact it purports to be. Rather, the "bare and blunt" social realism of the masses (as of *The New Masses*) was actually highly conventional. As Stevens wrote shortly after completing "Mr. Burnshaw and the Statue": "Poetry is like anything else; it cannot be made suddenly to drop all its rags and stand out naked, fully disclosed" (*L* 303).

In a similar vein, in "Owl's Clover" the Bulgar, whom Stevens identifies as "a worker," "a socialist" (*L* 371), is made to argue that the common man's desires are not addressed by art. "The workers," says the Bulgar, "do not rise, as Venus rose, / Out of a violet sea," but rather "inch / By inch, Sunday by Sunday" (*OP* 60). The phrase "the rise of the masses" sounds clearly in the background and plays against the imaginative and spiritual risings more often associated with Sundays in common usage and elsewhere in Stevens's poetry.[37] All risings are seen as fundamentally imaginative by Stevens. In contrast, the Bulgar who wants hands, noses, and eyes "massed for a head" champions a commonality consisting of unindividuated, purely physical characteristics. Stevens thus transposes the terms of his earlier argument with scientists, arguing that the "poet is individual. . . . The politician is general" (*L* 526).

Stevens mistrusted physical or materialist solutions—the section on the Bulgar is entitled "A Duck for Dinner" (*OP* 60)—and argued implicitly that the imagination underlies all political theorizing and that the processes of the imagination are more important than any particular theory. Tongue-in-cheek, he wrote to Latimer about Ruth Lechlitner's review: "We are all much disturbed about a possible attack from the Left; I expect the house to be burned down almost any moment" (*L* 313).[38] Stevens's response is amusing, of course, because he has taken his opponents at their word in their call for an overly simplistic tie between language and physical action in the world.

Other letters, as well as the sharp rebuttal of Burnshaw in "Owl's Clover," suggest that his critics had hit a sore spot. If Stevens was certain that his critics were on the wrong track, he was not certain that he had an adequate alternative. As he told Latimer, "Mr. Burnshaw applied the point of view of the practical Communist to IDEAS OF ORDER; in MR. BURNSHAW AND THE STATUE I have tried to reverse the process: that is to say, apply the point of view of a poet to Communism" (*L* 289). The same letter reveals Stevens's recognition that his argument might not suffice. He continues to say that the poem was meant to illustrate "the sort of contact that I make with normal ideas," and that he had "a good deal of trouble [in that] . . . it seems most un-Burnshawesque" (*L* 289). Similarly, he wrote to Ben Belitt, who criticized "Owl's Clover": "What I tried to do in OWL'S CLOVER was to dip aspects of the contemporaneous in the poetic. You seem to think that I have produced a lot of Easter eggs, and perhaps I have" (*L* 314). Stevens's search for a title for "Owl's Clover" also underlines his attempt to write poetry relevant to the contemporary world; his alternate title was "Aphorisms on Society" and his consideration of the title he finally chose included the admission that he wanted "to make poetry out of commonplaces: the day's news; and that surely is owl's clover" (*L* 311).[39]

In short, politics reactivated Stevens's anxiety that his celebration of process might have too little to do with commonplace living and might not adequately insist upon poetry's importance. The problem was one that had concerned him in one form or another since the 1890s. That is, Stevens responded to the pressure to defend poetry, first from the pragmatism of America in the 1890s and then again in the 1930s from the more literary pragmatism of the left. Both the 1890s and the 1930s were depression eras, of course, which increased the pressure to defend poetry as a worthwhile pursuit in the face of harsh economic reality. Even in better times, defenses of poetry in twentieth-century America required some convincing description of poetry's relevance to daily life. And the audience that needed convincing was not least the poets themselves.

Between 1924 and 1933 Stevens like Williams wrote relatively little poetry. By 1935, his letters suggest that he had been brooding about how to evolve a poetic style that would make his defense of poetry clearer both to himself and to others.[40] Although critical attacks on his first two volumes were responsible for his concern about reaching his audience, he was also in search of a theoretical foundation for his poetry that would enable him to continue defending his writing to himself.

The years in which Stevens wrote so little posed problems that were not merely theoretical. For one, he was busy establishing himself in the insurance business (*L* 243); he was also busy as a new father. Williams wrote to Pound, after they jointly had failed to elicit a poem from Stevens for a publication Pound was championing: Stevens "says he isn't writing any more. He has a daughter!"[41] Stevens himself wrote to Latimer in 1937 after Latimer's press folded: "Giving up The Alcestis Press must be to you what giving up any idea of writing poetry would be to me. Nevertheless, a good many years ago, when I really was a poet in the sense that I was all imagination, and so on, I deliberately gave up writing poetry because, much as I loved it, there were too many other things I wanted not to make an effort to have them. . . . I didn't for a moment like the idea of poverty" (*L* 320). There could hardly be a clearer statement of his belief that imagination and poetry were not easily related to practical, everyday life in America.[42]

Other letters also show Stevens's search for a convincing solution to the problem of how to relate poetry and reality. In 1937, while working on "The Man With The Blue Guitar," Stevens wrote that his stanzas dealt "with the relation or balance between imagined things and real things which . . . is a constant source of trouble to me. I don't feel that I have as yet nearly got to the end of the subject"; he goes on to claim that his poems "are not abstractions, even though what I have just said . . . suggests that" (*L* 316). Obviously Stevens worried that his poetry was not sufficiently linked to the real world in the sense that the poems might appear to others to be too abstract; in the sense that the theory informing the poetry was not fully worked out; and last, but not least, in the sense that his own life in the 1930s had made writing and living, or at least living well, difficult if not impossible to balance.[43]

These concrete concerns inform the poetry Stevens wrote in the 1930s. "The Man With The Blue Guitar" is unusual in the use of images drawn from contemporary American life.[44] The poem suggests that the earth needs poetry, without which it would be "flat and bare" (*CR*167). To offer human comfort, to give value to life, involves the attempt to "evolve a man," whose eye

> A-cock at the cross-piece on a pole
> Supporting heavy cables, slung
>
> Through Oxidia, banal suburb,
> One-half of all its installments paid.
> . . . .
> Oxidia is the soot of fire,
> Oxidia is Olympia.
>
> [*CP* 181–82]

The idea that language and poetry can replace the cross with the telephone pole, and turn Oxidia, "the typical industrial suburb" (*L* 790), into Olympia is not entirely convincing. More heartfelt is the poem's final assertion that "Monday's dirty light" (*CP* 183)—where "employer and employee contend" (*CP* 182)—is only cancelled when "we choose to play / The imagined pine, the imagined jay" (*CP* 184), that is, when imagination allows some escape from reality.

As early as 1909 Stevens voiced the idea to which he would return again and again, namely that reading (and writing) offered an escape from the world of commuters, office work, and, later, war (*SP* 205;

NA 30–36).[45] In the 1930s, however, his attempt to defend poetry was closely tied to his own and others' nervousness about how to address pressing social problems. Certainly, section thirty of "The Man With The Blue Guitar" (cited above) was an attempt to show poetry's relationship not just to the world but also to the contemporary American scene: Stevens's gloss talks of "the realism of a cross-piece on a pole" (*L* 362) as the focus of his imagined man of imagination, who is "an employe [*sic*] of the Oxidia Electric Light & Power Company" (*L* 791). The plea, as Stevens told Renato Poggioli, is to show how, "if Oxidia is the only possible Olympia, . . . then Oxidia is that from which Olympia must come" (*L* 788–89). Stevens set himself the problem of showing exactly how to find poetry in the suburbs of America, how stylistically and conceptually to relate imagination and reality in a more than abstract way both for himself and for his audience. "The Man With The Blue Guitar" insists there is a relationship, but it portrays more of an alternation or even antagonism between Oxidia and Olympia than any real vision of their interrelationship. The problem is one Stevens confronted again during the second world war, suggesting then that under such circumstances "why one should be writing about poetry at all is hard to understand" (*L* 501).[46]

Through most of the 1930s, Stevens continued to pit imagination against reason and to argue that the modern world's need for imaginative vision was precisely what made poetry timely. At the same time, he was clearly worried about how to insist upon the concrete importance of poetry in the contemporary world. He wrote, for example, that "Mozart, 1935," a poem he linked with "Owl's Clover" and that was clearly influenced by the Depression, was about "the status of the poet in a disturbed society, or, for that matter, in any society" (*L* 292). The same letter adds that there "is no reason why any poet should not have . . . status" (*L* 292) and that it "would be a great thing to change the status of the poet" (*L* 293).

During this period, apparently in order to give poetry and poets a solid status, Stevens toyed with the idea that a poet was a "biological mechanism, [producing what was] . . . beyond his power to change" (*OP* 220).[47] Thus Stevens comes close to Williams's description of the origin of a poem in "the very muscles and bones of the body" (*SSA*

98), although unlike Williams Stevens does not emphasize the technical skill involved in putting a poem on the page. The two men are most similar in that, just as with Williams's search for a new form in the 1930s, Stevens's redescriptions of poetry responded to contemporary pressures that forced him to return to the question of poetry's role in the modern world. His defense of the irrational was meant to reiterate his idea that poetry is not rational as philosophy and science are, and that it has a kind of necessity rather than being "absurd" (*L* 180) or a matter of "vanity" (*L* 131), as he had suggested earlier. During this same period, Stevens noted I. A. Richards's suggestion that the "Western ethic tends to find its sanction in biology," presumably because the idea echoed his own recent thinking.[48]

Yet to call writers "the merest biological mechanisms" (*L* 294) nearly overturns Stevens's point. Like Ananke, the figure of grim external necessity in "Owl's Clover," this flatly mechanical view hardly serves to distinguish poetry from science, nor to define poetry's place in the modern world.[49] That is, Stevens does not show why those who found no use for poetry should value it. And, as he wrote to Latimer in the mid-1930s, he felt there was a need for a "conception of the importance of [poetry]" (*L* 299).[50]

Stevens's strained attempt to articulate a conception of poetry's value that would convince both himself and others is seen most clearly in "Owl's Clover." In 1936, he said that he wrote "in order to formulate [his] ideas and to relate [him]self to the world" (*L* 306). Like "The Man With The Blue Guitar," "Owl's Clover" is clearly an attempt to reformulate his idea about poetry's relationship to the contemporary world, an attempt growing naturally out of his early poetics and the questions addressed there, although fueled by both personal pressures and by critical responses to his work.

A 1940 letter to Hi Simons discusses Stevens's desire to define poetry's centrality in the contemporary world. The mind "is a pretty hard thing to catch," Stevens wrote, an obvious reference to the mental or imaginative activity celebrated and embodied in the early poetry (*L* 352). He continued, saying he wanted "to get to the center," in part because he felt "isolated, and . . . wanted to share the common life" (*L* 352). A final note suggests again that his critics prompted some of these doubts: "People . . . say that I live in a world of my own; I think that I am perfectly normal, but I see that there is a center. For instance, a photograph of a lot of fat men and women in the woods, drinking beer and singing Hi-li Hi-lo convinces me that there is a normal that I ought to try to achieve" (*L* 352). The image, or photograph, of a beer party may suggest the normal, but it is a characteristic twist for Stevens to impart activity to the picture ("drinking" and "singing") and to include his own version of the drinkers' song.[51] In other words, it is inevitable that Stevens's contact with the normal should have appeared to him "un-Burnshawesque," both in that Stevens's central poetry had been defined precisely by its transformations of the normal, and in that he had typically refused to justify such transformations with the sorts of rational argument that might convince his critics. Although the proposed relationship between poetry and daily life was central to his early poetics, the embodiment of that relationship depended in part upon an acceptance of Stevens's playful tone.

Since Stevens typically avoids definitions, one would expect him deliberately and playfully to evade a direct response to the call for a socially and politically relevant poetry, such as he imagined Stanley Burnshaw and others demanding.[52] In fact, Stevens's letters at times do just this. He wrote (again to Latimer): "You will find occasional references in my poems to the normal. With me, how to write of the normal in a normal way is a problem which I have long since given up trying to solve, because I never feel that I am in the area of poetry until I am a little off the normal. The worst part of this aberration is that I am convinced that it is not an aberration" (*L* 287). As is argued most fully later, in "Imagination as Value," poetry's importance rests on what might be called constant aberration. In line with this, Stevens might have used "Owl's Clover" to toy with his opponents' point of view, just as in "A High-Toned Old Christian Woman" he appropriates and transfigures a religious vocabulary as much through controlled irony as through argument.

The transfiguration of a political vocabulary is the declared tactic in "Owl's Clover." In "Owl's Clover," however, Stevens's protagonists, whether Burnshaw or the generic Bulgar, are his real audience, so that his defense of poetry is not only intended seriously, it is equally intended to be taken seriously by a specific readership for whom, clearly, playfulness had not proved convincing.[53] In other words, there is a tension in "Owl's Clover" that Stevens treats with an uncharacteristic seriousness of tone, in that he knows that the commonplaces of the day, those dealing with political issues especially, do not have a real place in his poetry. As he wrote to Latimer: "Is poetry that is to have a contemporary significance merely to be a collection of contemporary images, or is it actually to deal with the commonplace of the day? I think the latter, but the result seems rather boring" (*L* 308).

Ironically, Stevens comes close to portraying himself as Williams had described him almost twenty years earlier in the "Prologue" to *Kora in Hell*, where Williams quotes Stevens arguing for sticking to one's subject and against "fidget[ing] with points of view" (*I* 15). In contrast, Williams finds "attention has been held too rigid," and advocates loosening the attention (*I* 14). The disagreement between the two poets involved some willful misunderstandings. After all, both agreed that the "imagination goes from one thing to another" (*I* 14), even though Williams states this credo as if Stevens would disagree, and quotes excerpts from a letter by Stevens that repress the tensions Stevens usually celebrated. The clash recalls Stevens's famous—or infamous—description of Williams's poetry as "anti-poetic" and of Williams as "more of a realist than is commonly true in the case of a poet" (*OP* 255–56). Here too both men agreed that poetry centrally involved what was real; Stevens's adjective, "anti-poetic," is not uncomplimentary. Yet given Stevens's resistance to identifying the real with local detail, his description of Williams as anti-poetic was, as Williams saw, double-edged.

The arguments between Stevens and Williams reveal, in one sense, some genuine disagreements, especially on the issue of how and whether poetry could represent quotidian details of the contemporary world. In another sense, however, both poets were arguing with different sides of themselves. This is also apparent in Stevens's argument with Burnshaw. Stevens knew that his usual stance, that political visions are fundamentally imaginative visions, was most compelling where he avoided, and therefore undermined, the format of rational, political discourse. His overreaction to Burnshaw stemmed from his recognition that in order to be taken seriously, he would have to confront the discrepancy between his essential imagination and a duck for dinner—between poetry and the practical world—and to answer his critics he tried to do so on a level of discourse where rhetorical evasions might seem less a celebration of human imagination than an abandonment of basic human needs.

Stevens's major recognition in "Owl's Clover" is documented in his return to a more characteristic style in the final sections of the poem. The overall movement of these sections, from the creation of the myth-like portent to its dismantling in the final return to the "hum-drum" (*OP* 71), represents the cyclic activity of the imagination.[54] Stevens says that the hum-drum is "indifferent to the poet's hum" (*OP* 71) yet implies by his choice of language that there is even so a "hum" in "hum-drum." That is, both poetic creation and the denial of poetry are imaginative processes.

To underline these processes, and to focus attention on the human importance of poetry, the poem offers a series of human figures, such as the orator, the sculptor, and the subman, that are both emblems and products of the activity of figuration on which Stevens's style insists. No one static figure can adequately represent poetic process: the activity by which man might recreate his image of himself and his world can only be continual if the images proposed are dismantled in order to allow for new images. In fact, the dismantling of his figures is what distinguishes Stevens's central imagination from the ideology he opposes in "Owl's Clover." Insofar as his central imagination cannot be portrayed in a static figure, the figures proposed allow Stevens to make his grand claims for art and imagination because they are qualified and replaced.

Stevens's attempt to meet his critics on their own ground, by dealing with the day's news, as he put it, reveals both the strengths and weaknesses of the poetry he defends. Poetry after all cannot meet certain kinds of practical demands, and the issue is one Stevens would not argue again in the same way. Williams's 1934 poem "The Death of See" (*CEP* 381–82) more literally dealt with the daily news by resisting its language. In his poem, Williams argued with newspaper accounts of Harry Crosby's death, taking issue both with the opaque language and the image of poets found in the papers.[55] Stevens, on the other hand, was responding to critics who found his work politically irrelevant. Still, and in spite of the obvious differences between the two poems, both imply that poetry's strength lies in its ability to engage the mind and to see through clichés and dead language. In Stevens's case, his argument with his critics helped him to clarify his ideas. In the process of delineating the need for a contemporary imagination, he found a way of emphasizing and reaffirming the activity that made poetry central for him, both in his style and in his use of figures to call attention to the human, social, and political importance of the poetics of process. Later, armed with a new view of poetry's centrality, Stevens could say (without feeling the need to discuss ducks for dinner) that his attempt was "to try to get as close to the ordinary, the common-place and the ugly as it is possible for a poet to get" (*L* 636).

In the 1930s, then, Stevens evolved many of the elements of his late style, in part as a way of emphasizing imaginative process while still providing images that momentarily fix that process in order to underline it. There is much more than mere humor involved in the punning by which figuration is emblematized by human figures, since these figures exemplify the human activity out of which they arise, namely the use of language and the celebration of its excesses. If such strategies did not convince Stevens's critics, to some extent they satisfied his own urge for "sovereigns of the soul" (*CP* 124).

While "Owl's Clover" is important in the development of Stevens's late style, his most assured account of why imaginative process should be valued appeared a few years later, in the 1940s.[56] Stevens said, in fact, that he accepted the invitations to give the papers in which his poetics are most clearly articulated because they made "it necessary to take a good look at ideas that otherwise would drift about, vaguely."[57] In these essays, he reviewed his claims for poetry and went on to point out how poetry was also practical—that is, linked to objective reality—because it expressed and even anticipated the results of modern physics. Given this use of physics, it is not surprising to find one of Stevens's colleagues noting of Stevens at business meetings: "When we bordered on some scientific development, . . . he was interested."[58] The ability to show how the world celebrated by poetry was the same world arrived at by physics proved the defense of poetry's value for which Stevens was looking; it was a way "to create a perspective for poetry: that is to say, to give it a hearing and a position" (*L* 382).

Wallace Stevens: A Supreme Hesitation

Much of Stevens's knowledge of modern physics came from friends or from articles found in popular magazines. He referred to the new physics as early as 1941, but a 1951 letter to Barbara Church suggests that it was not until he received Jean Paulhan's letters on modern philosophers of science and quantum theory that Stevens really registered how thoroughly he might use those fields (*L* 725). Stevens's career may also have aided in his understanding of modern physics. If Stevens was a failed reporter who disliked and mistrusted naturalism, the management of surety bonds must have given him some insight into modern physics, with its statistical truths that are universally true and yet always leave knowledge of specific cases incomplete. Certainly his essays from the 1940s and 1950s increasingly repeat the affinity he felt between his view of poetry and quantum theory.

Stevens's early use of the new physics is exemplified in his essay, "The Noble Rider and the Sound of Words," written in 1941. Allen Tate's preface to the volume in which Stevens's essay was published says that all the essays have as a common premise "that poetry, although it is not science, is not nonsense," an indication of the generally shared perception that poetry needed a new defense in the modern, scientific world.[59] I. A. Richards's essay, "The Interaction of Words," may bear out Tate's introductory remarks most clearly. Richards discusses how science underwrites poetic truths, since for modern poets "the distinction between matter and activity vanishes—as it does for the modern physicist when his ultimate particles become merely what they do" (p. 84).

This suggestion was an important one for Stevens, although Richards's essay may not be where he first encountered the idea. Stevens's early volumes had insisted upon how perception must be a constant activity, which gives us our sense of the world and by which we constitute the world. Many of the poems celebrate the invigorating exchange between self and world, yet others mourn the loss of static, certain objects that might be known once and for all. "The Snow

Man," for example, may end by suggesting that abstractions ("the nothing that is") best provide naked reality, but the poem also voices a desire for a reality untainted by human intellect or imagination. "The Man on the Dump" also notes how any given image of the world must be discarded and concludes with a nostalgia for the truth—"The the" (*CP* 203), that is, for some solid presence, even as it, like most of Stevens's poetry, presents the "world [as] a force, not a presence" (*OP* 172). Even when Stevens sounded most at home with unending, provisional encounters with the world, he voiced anxiety about how the importance of such activity, exhilarating as it might be, could be defended. In particular, Stevens's statements from the 1930s suggest his worry that the motions he celebrated and the uncertainty he courted never adequately presented the *central* of self or world.

The dissolving figures that appear in "Owl's Clover" offered a stylistic solution to the problem of how to emphasize the human necessity and relevance of a process that can never finally get the world right. Richards's notes on modern physics suggested that such radical uncertainty had a basis in fact. If the ultimate building blocks of the universe are simply what they do, if they, like Stevens's creative mind, cannot be fixed, then the elusive truths that poetry enacts have a solid, scientific foundation. Although the interaction between the world and the perceiving imagination long preoccupied Stevens, earlier he could not theoretically reconcile the uncertain world thus uncovered with the reality of industrial suburbs, economic hardship, and everyday living. In "The Noble Rider," however, we find an appeal to philosophy and physics, which had certified that vibrations, movements, and changes, always central to Stevens, were indeed "things as they are."

Russell's 1941 *Let the People Think* also notes the effect that modern philosophy, from the American pragmatists to Bergson, and the new physics had on notions of truth, mentioning both Bergson and Whitehead in one breath.[60] Stevens cites Russell's book in "The Noble Rider," when he talks about the importance of Americans acquiring immunity to eloquence. The context on which Stevens draws helps show how the development of his poetics was in part a response to typically American attitudes toward, as well as respect for, science.

In his essay, Stevens considers several sculptures and public images that, like the statues in "Owl's Clover," either fail to emphasize imagination or, conversely, are too detached from contemporary reality. Stevens rejects as mere fancy the statue of an American hero, Andrew Jackson, in favor of a painting of a carrousel.[61] It is not surprising that Stevens liked the motions and joie de vivre underlined by a merry-go-round, the very name of which emphasizes the pleasures of merely circulating, even as the image is drawn from contemporary American life. It is also characteristic of Stevens to prefer figures on wooden horses that move—and go nowhere in particular, moving for the pleasure of movement—over public figures on static, monumental stone horses.

It is worth asking, however, why Russell is mentioned in this discussion of American images. Russell's book helps clarify the point; he asks there for an emphasis upon the process of scientific thinking as opposed to the products of science (p. 42). Moreover, Russell contrasts a Renaissance joie de vivre, echoing a Stevensian phrase (see *L* 793), and the American association of science with practicality and utilitarian language (pp. 81–85). In his essay, Stevens is unhappy about the aspect of the American character displayed in Clark Mills's statue, which draws on commonplace associations rather than the mind's free play. It is a work that shows selections "made for purposes which are not then and therein being shaped but have been already fixed" (NA 10–11). Here, in contrast to the defensive treatment of ideology and politics in "Owl's Clover," Stevens gives his most convincing argument as to the central place of poetry in a democratic society. Like Russell, Stevens celebrates the joy of language and thought in process rather than language, thought, or image in the service of a predetermined practical end. Yet Russell insists that scientific progress has grown out of just such mental play, adding weight to the enterprise poetry valorizes. In ways, Stevens overturns Russell's proposed stereotype of Americans as relentlessly practical. At the same time, however, Stevens allies his playful style with the cast of mind that produced tangible scientific progress, and so implies his very impracticality is practical.

By the 1940s, Stevens's claim, like Williams's, was also that his poetic style could provide the language required by modern physics even as he invoked physics to sanction modern poetry. Again in "The Noble Rider," Stevens quotes a 1941 article by C. E. M. Joad from the *New Statesman*, in which Joad discusses how philosophy and modern physics have "dismissed the notion of substance." Perception, Joad says, is constantly changing and similarly "with external things. Every body, every quality of a body resolves itself into an enormous number of vibrations, movements, changes. What is it that vibrates, moves, is changed? There is no answer" (*NA* 25).[62] Stevens adds that the poet is not interested in the intellect's false view of a " 'collection of solid, static objects extended in space' but the life that is lived in the scene that it composes; and so reality is not that external scene but the life that is lived in it. Reality is things as they are" (*NA* 25). Here Stevens discovered the definition of reality that had eluded him earlier.

After writing "Owl's Clover," and perfecting his use of dissolving figures to underline the role of the imagination without abdicating his insistence upon motion or process, Stevens wrote to Hi Simons: "Logically, I ought to believe in essential imagination, but that has its difficulties. It is easier to believe in a thing created by the imagination. A good deal of my poetry recently has concerned an identity for that thing" (*L* 370). In spite of his insistence on forces as opposed to presences, Stevens realized he required some entity that might represent the process he celebrated as documented in "Owl's Clover." Joad's statement about physics—what "is it that vibrates, moves, is changed"—proposes how language might capture "things as they are," given that there are no things, in the sense of "solid, static objects." Stevens, who always found that ambiguity fueled the imaginative play of language, could thus claim he had evolved the poetic style required by the world Joad describes, even as he implied that physics guaranteed modern poetry's relationship to reality.

There is, according to physics, no *it*, and the very syntax of Stevens's sentences seizes upon this revelation. He discusses nobility saying, "I am evading a definition. If it is defined, it will be fixed and it must not be fixed. As in the case of an external thing, nobility resolves itself into an enormous number of vibrations, movements, changes. To fix it is to put an end to it. Let me show it to you unfixed" (*NA* 34). The essay ends with a dazzling string of anecdotes and associations that illustrate the play of the mind, the process, which cannot be fixed or encapsulated. At the same time, as with his use of the figures and aphorisms that punctuate the poems, Stevens provides a momentary and qualified fixity, a way of emphasizing his point without destroying it, in his use of a seemingly logical introductory paragraph ("Let me show it to you unfixed") and his use of a pronoun (it), the reference for which scientists agree cannot be found.[63] Again, poetry provides the language physics needs even as physics is placed in the service of poetry in that Stevens uses it to underwrite a long-standing practice.[64]

In 1945, Stevens could say that he was still engaged "in defining the place of poetry" (*L* 500–01). Such disclaimers, however, do not negate the increased assurance with which Stevens treats his defense of poetry and its ties to reality in essays such as "The Noble Rider" or, from 1948, "Imagination As Value." The source of Stevens's assurance in such essays is related to his discovery that the imagination he celebrated, and the style in which he presented that celebration, could be seen as part of objective reality, given his understanding of modern philosophers and physicists as gleaned from Joad, Russell, and others. Just as the figure of Max Planck closes the 1951 "A Collect of Philosophy," "Imagination As Value" ends with the figure of Jean Paulhan, one of the men who later wrote to Stevens explaining quantum physics (*OP* 195–96, 200) and noting the physicists' need for the poets' language: "Progress in physics is . . . in suspense because we do not have the words or the images that are essential to us" (*OP* 196).[65] By the 1950s, in part because of Paulhan's letters, Stevens was able to draw on a detailed knowledge of the new physics; he also frequently cites sources such as Whitehead.[66]

Stevens did write Sister Bernetta Quinn that James and Royce rather than Whitehead were the philosophers to whom he had been exposed while in college, and that philosophy of science was not a subject he had studied. Yet it is hard to believe his disclaimer that in 1950 he had still not read Whitehead at least as seriously as he read anything outside of poetry (*L* 704), especially in light of his extensive use of the philosopher in his 1955 "The Whole Man: Perspectives, Horizons" (*OP* 229–35). Even Stevens's 1944 protest to Paul Weiss, calling Whitehead "purely academic" and tacitly accepting Weiss's suggestion that he had been reading only the theories of James and

Bergson (*L* 476), is curious in light of popular conceptions of James and Bergson as precursors of the new physics. Stevens must, for instance, have read Williams's 1946 "Choral: the Pink Church," which linked objects in the world and the light in which they were seen ("Sing! / transparent to the light . . . /—that is the light and is / a stone / and is a church" [*CLP* 159–60]) calling James, Dewey, and Whitehead the saints of this new vision (*CLP* 161). In spite of Stevens's ambivalence about Williams's work, surely he would have been captivated by Williams's suggestion in this poem that poetry might replace religion and would have pursued Williams's invocation of Whitehead along with philosophers whose work Stevens said he did know and admire. Moreover, in the final chapter of *Science and the Modern World*, Whitehead, like Dewey, emphasizes the importance of "creative activity" and the dangers of disembodied knowledge, ideas that Stevens as well as Williams would have found attractive.[67] In any event, by the early 1950s, Stevens's essays contain quotations from Whitehead.

To some degree, Stevens's later references to science simply extend the way he was already referring to science in essays such as "The Noble Rider." The 1951 "A Collect of Philosophy," for example, repeats the idea that physics requires the language and images of poetry (*OP* 196) and quotes Whitehead as follows: "My theory involves the entire abandonment of the notion that simple location is the primary way in which things are involved in space-time. In a certain sense, everything is everywhere at all times, for every location involves an aspect of itself in every other location. Thus every spatio-temporal standpoint mirrors the world" (*OP* 192). Stevens comments: "These words are . . . from a level where everything is poetic, as if the statement . . . produced in the imagination a universal iridescence, a dithering of presences and, say, a complex of differences" (*OP* 192). By adding his own typical vocabulary for mental exhilaration to Whitehead's universe, Stevens again appropriates science as the basis for the poetry he had long defined as involving interrelationships and uncertain presences; citing Whitehead, he could claim without qualms that such an ambiguous world was both real and objective.

Whitehead also introduced quantum theory and proposed that the scientific materialism associated with nineteenth-century science could be challenged, noting that "each primordial element [is] a vibratory ebb and flow of an underlying energy, or activity."[68] Having "associate[d] the quantum theory with vibration" (p. 52), Whitehead went on to argue for a "new doctrine of organism" (p. 53). Perhaps most importantly, Whitehead linked his theory with Romantic poetry, which, he said in the chapter that Stevens cited, stood "against the exclusion of value from the essence of matter of fact" (p. 133). In sum, Whitehead proposed that one could be an objectivist without being a materialist (p. 127).

Whitehead further suggested that modern philosophy could heal the post-Cartesian split between mind and matter by bringing "together the two streams into an expression of the world-picture derived from science, and thereby end the divorce of science from the affirmations of our aesthetic and ethical experiences" (p. 218). In 1953, Stevens looked back at section twenty-five of "The Man With The Blue Guitar" (involving a man who balances the world on his nose, keeping it in motion) and said the man was any "observer: Copernicus, Columbus, Professor Whitehead, myself" (*L* 790). The choice of Whitehead is not casual. Again, the balancing act that had been difficult to defend in the 1930s is here seen as the real world, uncovered by poets and readers, but also discovered and validated by science and philosophy. The company in which Stevens places himself (Copernicus, Columbus, Whitehead) links poetic discovery with the tangible or verifiable discoveries of the world by literal voyagers and scientists. Thus scientists or philosophers of science, Whitehead in particular, ratified Stevens's desire to include reality in poetry without identifying reality with material reality or objects. Such a rationale allowed Stevens to write that the "ultimate value is reality" (*OP* 166) and the "final poem will be the poem of fact in the language of fact" (*OP* 164), while still writing and defending his poetry of process.

In short, Stevens used modern science and explicators of modern science (like Joad and Whitehead) to develop theoretical foundations for his poetry. As he wrote ten years after his period of silence, "If you don't believe in poetry, you cannot write it" (*L* 500). He also wrote that he needed "a true belief" (*L* 793). Modern science allowed Stevens to demonstrate the universality, the truth, of his belief in process.

He maintained his insistence that poetry involved "the joy of language," but he called on physics to testify that the unending process of describing and redescribing parts of the world, and thus one's encounters with the world, in language was the most accurate picture of reality available.[69] His use of Max Planck at the end of "A Collect of Philosophy" is exemplary. Paulhan's letters and André George's note on Planck in *Les Nouvelles Littéraires* helped Stevens add a new dimension to his attempt, as he put it, to show how poetry did "matter nowadays" (*L* 729).[70]

"A Collect of Philosophy," typically, proceeds by digressions, illustrations, and examples (many of which are human figures) to illustrate what Stevens calls excess, or verbal play. As first explored in "Owl's Clover," it is as much the way the essay proceeds as the particular figures that are important. Nonetheless, the choice of Planck is not accidental. Stevens concludes his essay "by placing here at the end a figure which would leave the question of supremacy [between imagination and reason; poetry and philosophy] a question too difficult to attempt to solve" (*OP* 201). Planck, Stevens writes, is more appropriate than Pascal because at the end of his career Planck displayed what George called "a supreme hesitation" (*OP* 201). Stevens characterizes this hesitation as "a nuance of the imagination" (*OP* 202).

Planck's indecision has to do with, among other things, both the origins of his discoveries and with the nature of those discoveries. Technically, Planck's 1900 paper, "On an Improvement of Wien's Radiation Law," marked the beginnings of quantum mechanics. In that paper, Planck found that if scientists replaced the classical notion of a continuum of energies with a quantized *ladder* of energies, a problem in classical physics known as the "Ultraviolet Catastrophe" could be solved. Roughly speaking, Planck's discovery allowed that red-hot chambers contained—as in classical physics—energies of arbitrary value, but that the sum total of energy was not infinite, as classical physicists uncomfortably had to conclude. Although Planck's constant (*h*) is to this day believed to govern the sizes of quantum steps, there were inconsistencies in his arguments and his work paved the way for, among other things, Heisenberg's Uncertainty Principle, narrowly a theory about error in position and momentum measurements of waves, which nonetheless led Heisenberg himself to announce that "a path [of a particle] comes into existence only when we observe it."[71]

Stevens quotes not only George on Planck but also Paulhan's explanation of Planck's discovery, namely that "the true nature of corpuscular or quantic phenomena" cannot be known (*OP* 195). Both the origins and the nature of Planck's discovery, therefore, aptly parallel Stevens's point, that the imagination is set in motion by questions that cannot be resolved. Furthermore, the way in which Stevens arrives at his final figure, giving us Pascal and then replacing him with Planck, is also an embodiment of the hesitation he wishes to emphasize. In fact, by letting the suggestion of Pascal as a possible figure remain in the essay despite his announced change of mind, Stevens leaves the reader with both figures, thus providing another instance of the excess of poetry on which he focuses earlier in the essay, an excess now allied with scientific theories about the physical world.

Excess and fluctuation were not new features of Stevens's poetics. Even in the 1930s, when he played with the idea of the poet as a biological mechanism, he wrote: "where there are no fluctuations, poetic energy is absent" (*OP* 229). Stevens's new understanding of Planck, however, let him develop his ideas on poetic fluctuation and energy. Not only did Planck's work certify that fluctuation was part of the world, but, as George wrote, Planck himself, who had been a determinist, was bothered by the universe of probabilities rather than of determined, causal phenomena that he had discovered. The hesitation Stevens cites is Planck's as well as the world's, although George's prose makes it difficult to tell which hesitation is at issue—that of Planck in accepting his results or that inherent in his theory, which made causality "neither true nor false" (*OP* 202). George's linguistic ambiguity exactly parallels both hesitations. In the end, Stevens appropriates for poetry the ambiguous relationship between Planck, the language in which George described him, and the relationship Planck proposed between man, language, and the world.

The figures of Pascal and Planck are specifically raised in the course of asking whether philosophy or poetry is greater.[72] Stevens's suggestion that philosophers and philosophy, physics and physicists, are poetic shows that poetry and imagination are central—not mere play, but part of endeavors often viewed, especially in America, as more valuable than poetry and as providing greater access to truths about the world. As Stevens explained: "To say that philosophers are poets . . . does them no harm and at the same time . . . magnifies poetry" (*L* 734). Further, as in Williams's *Embodiment of Knowledge*, poetry is said to subsume science and philosophy, taking over the task Whitehead had marked out for philosophers, namely remarrying the mind, formerly the province of philosophers, and the world, formerly the province of scientists. Even as poetry is elevated in this way, science and philosophy are asked to verify that modern poetry is needed to provide some means of describing reality. When Stevens leaves us with both figures, moreover, he suggests poetry's inclusiveness, its unique ability to present a kind of Coleridgean unity in diversity. As early as 1940, trying to convince Henry Church to establish a chair of poetry, Stevens wrote that the "knowledge of poetry is a part of philosophy, and a part of science" (*L* 378). By 1951, philosophy and science are parts of poetry. As in "Notes Toward A Supreme Fiction," poetry is proposed as the supreme activity, in that it can appropriate and relate both philosophy and physics: "It was not a choice / Between, but of. He chose to include the things / That in each other are included, the whole, / The complicate, the amassing harmony" (*CP* 403).

Stevens's philosophical affinities have often been discussed. Less commented on is how frequently he drew on scientific theories to bolster and shape his poetics. One can say, moreover, that his use of science in the search for an adequate defense of, and figure for, his essential imagination was in turn shaped and necessitated by the American context, its love of the tangible, its respect for hard fact, its suspicions of poetry. Stevens's development is not easily explained by pointing to what influenced him. To speak of his confusing juvenile discussions of poetry's place or of how poetry differs from science is not to explain the emergence of the poems found in *Harmonium*. Nor does showing his recognition of how poetry could be affirmed by modern physics in itself explicate the later poetry. Even if Stevens could continue his defense of poetry in part because of his use of modern physics, he did not stop asking questions about whether poetry was important. One of his last poems still asks: "I wonder, have I lived a skeleton's life, / As a disbeliever in reality" (*Palm* 396)?

"As You Leave the Room" also includes a self description of Stevens as a "countryman of all the bones in the world" (*Palm* 396). To place Stevens in a specifically American context, however, is to help show how even one of the apparently most abstract of the American modernists was deeply influenced by the climate in which he wrote. Knowing American attitudes toward poetry and science helps clarify Stevens's felt need for a defense of poetry's place in the world and his preoccupation with finding that poetry was tied to what he could convincingly call objective reality or fact.

Stevens himself suggested that America presented its poets with particular difficulties. It is no accident that one of the poems written during the 1930s, Stevens's period of highest anxiety over whether poetry could be shown to be important, specifically invokes America. "The American Sublime" asks: "How does one stand . . . To confront the mockers, / The mickey mockers" (*CP* 130). The poem rejects as inadequate a statue of Andrew Jackson, which "The Noble Rider" later identifies as a reference to a monument found in Washington, D.C., that is, as an attempt to present an American ideal. The poem ends with questions about how to mark secular communions of "spirit and space": "What wine does one drink? / What bread does one eat?" (*CP* 131). I have tried to argue that not only the shape of Stevens's evolving answer to that question, including his attention to the discoveries and characterizations of science, but also the very seriousness with which he asks are indicative of his need to describe poetry's importance in modern America.

The Speech of the Place

If Stevens did not incorporate the commonplace objects of American life into his poems or compare his poetry to such objects, he nonetheless developed his mature poetry and his influential poetics in part because of the need that he felt for a poetry that offered people something more vital than public statues of American heroes. His transition from the earlier, more anecdotal poems to the later, more meditative major poetry was prompted by his need to reconcile an urge for order with a poetics of process, a reconciliation that was occasioned, in turn, by the pressure to find a description of poetry's importance that would sound convincing to an American audience. All three of the poets examined here can be said to have responded to this same pressure and, with varying emphases, to have reconciled two poetics in the end by balancing their commitments to objects and observers, artifacts and energy, the creative process of the poet and the public role of poetry.

All three poets also attempted to draw on analogies between poetry and science, technology, or business to explain the importance of poetry to the American public. Moore invited comparisons between poems and other objects in the world, from snakes to machines, and, in a more encompassing gesture, she insisted on the relationship between poetic and other kinds of productive or creative energy. Her defenses of poetry strategically appropriate spheres other than the poetic; moreover, the poems themselves use and redefine the vocabulary of public life (of science, business, and advertising) in perhaps her most convincing demonstrations of poetry's participation in and enrichment of a shared, public realm: that of language. Williams, too, defended poems by appeal to American productivity in other areas, and finally, like Stevens, rested his case for poetry's importance on the apparently scientific nature of the reality poetry captured and enacted. In their poems, Williams and Stevens, like Moore, turn to language as part of poetry's authority in a public realm.

Stevens's figures for poetic activity generally read, speak, or write, emphasizing the linguistic activity by which they—and other, more obviously public, figures—are created. "Country Words," for example, begins with "Belshazzar, putrid rock" and through the words of a song (the poem) sung by the narrator, Belshazzar is transformed to one "reading right / The luminous pages on his knee, / Of being" (*CP* 207). In effect, the narrator's desire and resultant activity create the figure desired, so that the final image of Belshazzar reading and alive is no longer that of a pillar "of a putrid people" (*CP* 207), but of a poetic process that might save societies from putrefying. Similarly, "Examination of the Hero in a Time of War" transforms a central figure of a hero-skeleton, calling attention to the activity by which such figures are created and changed. The poem ends with the "highest man" grown to "man-sun, man-moon, man-earth, man-ocean," and then brought back down to earth like the portent in "Owl's Clover": "The man-sun being hero rejects that / False empire" and arrives "at the man-man as he wanted" (*CP* 280). The exaggerations of the public hero as more than human are rejected. Significantly, though, it is one of the exaggerated figures that proves capable of rejecting itself. Attention is focused, thus, on the creation and dismantling of figures, that is, on the poetic process by which public values are formed and changed.[73]

Williams's attention to poetry's ability to provide and maintain the health of public ideals is even more explicit. In "To Elsie," he underlines the need for someone "to witness" (*CEP* 272), to express the lives of the "pure products of America" (*CEP* 270). And *Paterson* opens with a search for a "common language" (*Pat* 7): "The language, the language / fails them / They do not know the words / or have not / the courage to use them" (*Pat* 11). The poem maintains this search, ending with Paterson's effort "to get the young / to foreshorten / their errors in the use of words" (*Pat* 230–31). If Williams knew that his language sometimes missed the audience he hoped to reach, his work makes clear that the inability to hear poetry is part of "the truth about us" (*I* 132) and by bearing witness to that truth in his poetry, Williams helped to forge the language he thought was needed.

Williams included in *Paterson* an interview where he was repeatedly asked "is it poetry?" (*Pat* 224–25). Citing one of Williams's lesser poems and pairing it with one of e. e. cummings's least accessible pieces, the interviewer (Mike Wallace) baits Williams. He proposes, for example, that Williams's poem is only "a fashionable grocery list" (*Pat* 224). Wallace's tone and choice of poems suggest his antagonism toward modern poetry. By including the interview in *Paterson*, Williams acknowledges that the poets' examination and revaluation of public language did not necessarily change American attitudes toward poetry. Neither did the poetic defenses that placed poems on a par with industrial products or compared poets with scientists and engineers. Yet if the modernists recognized the limits within which they worked, they achieved a good deal within these limits.

Stevens's modern poetry had "to learn the speech of the place. / It ha[d] to face the men of the time and to meet / The women of the time . . . to think about war" (*CP* 240) even while composing its own stage and audience. All three poets discussed here might be said to have followed Stevens's injunction. *Paterson*, for example, addresses itself specifically to the problem of war and to learning "the speech of the place." Further, Williams includes not only Paul Klee, Freud, Picasso, and Juan Gris, but, in their midst, Mike Wallace. If relatively few of the men and women of the time understood what the poets were attempting, the American modernists examined here were astute in asking why this was the case and in diagnosing some of the problems poetry faced in America. As Williams wrote to Moore, "At times there is no other way to assert the truth than by stating our failure to achieve it. If I did not achieve a language I at least stated what I would not say" (*SL* 304).

Stevens's double aim of facing the modern world and of defining poetry's place in it, Moore's meditations on American values and education, and Williams's search for an American language that Americans would read not only participate in a tradition; they also form our tradition and underlie current struggles to define the value of being a poet in this country.

Notes

INTRODUCTION

1 J. Meredith Neil, *Toward a National Taste: America's Quest for Aesthetic Independence* (Honolulu: University Press of Hawaii, 1975); Alan Trachtenberg, *The Incorporation of America: Culture and Society in the Gilded Age* (New York: Hill and Wang, 1982); and Jackson Lears, *No Place of Grace: Antimodernism and the Transformation of American Culture, 1880–1920* (New York: Pantheon Books, 1981). See also David Perkins, *A History of Modern Poetry from the 1890's to the High Modernist Mode* (Cambridge: Harvard University Press, 1976); Richard Hofstadter, *Anti-Intellectualism in American Life* (New York: Vintage Books, 1963); and John Tomsich, *A Genteel Endeavor: American Culture and Politics in the Gilded Age* (Stanford, California: Stanford University Press, 1971).

2 See John Passmore, "Logical Positivism," and Nicola Abbagnano, "Positivism," in *The Encyclopedia of Philosophy*, ed. Paul Edwards, vol. 6 (New York: Macmillan, 1967), pp. 52–57, 414–19, and Charles Morris, *The Pragmatic Movement in American Philosophy* (New York: George Braziller, 1970). See also E. A. Burtt, *The Metaphysical Foundations of Modern Physical Science* (rev. ed. 1932; reprint ed., Garden City: Doubleday–Anchor Books, 1954) and Leszek Kołakowski, *The Alienation of Reason: A History of Positivist Thought*, trans. Norbert Guterman (Garden City: Doubleday–Anchor Books, 1968).

3 "Things That Have Moulded Me," *Dial* 83(September 1927):184. See also Russell's *Icarus; or The Future of Science* (London: Kegan Paul, Trench, Trubner, and Company, 1924), a book found in Stevens's library, for a similar view.

4 Howard P. Segal, *Technological Utopianism in American Culture* (Chicago: University of Chicago Press, 1985), pp. 12–13.

5 *Technological Utopianism*, p. 13.

6 See also *Technological Utopianism*, p. 14: "If technology today is increasingly based upon scientific principles, and if scientific research today is as much applied as 'pure,' this interdependence is a relatively recent development." Further discussion of the interrelations of science and technology can be found in Ihab Hassan, *The Right Promethean Fire: Imagination, Science, and Cultural Change* (Urbana: University of Illinois Press, 1980) and Douglas Davis, *Art and the future; A History/Prophecy of the Collaboration between Science, Technology, and Art* (New York: Praeger, 1973). Finally, see Martin Heidegger's 1954 essay on "The Question Concerning Technology," reprinted in Martin Heidegger, *Basic Writings*, ed. David Farrell Krell (New York: Harper and Row, 1977), pp. 287–317.

7 For more detailed discussions of modernism, see Dickran Tashjian, *Skyscraper Primitives: Dada and the American Avant-Garde 1910–1925* (Middletown, Connecticut: Wesleyan University Press, 1975); Hugh Kenner, *The Pound Era* (Berkeley and Los Angeles: University of California Press, 1971); Malcolm Bradbury and James McFarlane, "The Name and Nature of Modernism," in *Modernism*, ed. Bradbury and McFarlane (Harmondsworth, England: Penguin Books, 1976); Peter Faulkner, *Modernism: The Critical Idiom* (London: Methuen, 1977); Monroe K. Spears, *Dionysus And The City: Modernism in Twentieth-Century Poetry* (New York: Oxford University Press, 1970); Robert Rosenblum, *Cubism and Twentieth-Century Art* (1959; rev. ed. New York: Harry N. Abrams, 1976); Joshua C. Taylor, *Futurism* (New York: Museum of Modern Art, 1961); Max Kozloff, *Cubism/Futurism* (1973; reprint ed., New York: Harper and Row, 1974); William S. Rubin, *Dada and Surrealist Art* (New York: Harry N. Abrams, 1968); Hans Richter, *Dada: Art and Anti-Art* (London: Thames & Hudson, 1965); Meyer Schapiro, *Modern Art: 19th and 20th Centuries* (New York: George Braziller, 1978); and Henry Russell Hitchcock's introductions in *Modern Architecture International Exhibition* (1932; reprint ed., New York: Arno Press for The Museum of Modern Art, 1969).

8 John A. Kouwenhoven, *The Arts in Modern American Civilization* (1948; reprint ed., New York: W. W. Norton, 1967), pp. 13–42.

9 Some critics have suggested calling this revolution a paradigm shift, in

Thomas Kuhn's sense of the word, paralleling or related to the paradigm shift in the visual and literary arts (Thomas S. Kuhn, *The Structure of Scientific Revolutions* [Chicago: University of Chicago Press, 1962]). I, however, have avoided the word *paradigm*, which implies a view of scientific change about which there is significant disagreement among historians of science. Since my topic is popular views of science in a given period, it did not seem wise to enter into debates about the actual nature of changes in scientific fields. Moreover, Kuhn himself has disclaimed attempts to apply his theories to the field of art, in *The Essential Tension: Selected Studies in Scientific Tradition and Change* (Chicago: University of Chicago Press, 1977), especially in "Comment on the Relations of Science and Art," pp. 340–51.

10 Alan J. Friedman and Carol C. Donley, *Einstein as Myth and Muse* (Cambridge: Cambridge University Press, 1985), pp. 10–20.

11 I am grateful to Dr. Richard Crandall of the Reed College Department of Physics for his willingness to read this and offer suggestions. For more technical descriptions of Einstein's findings, see Max Jammer, *Concepts of Space* (Cambridge: Harvard University Press, 1954). Donley and Friedman, on the other hand, offer an accurate but less technical explanation of Einsteinian physics in *Myth and Muse*, pp. 45–66.

12 See *Myth and Muse*, pp. 110–17, and Max Jammer, *The Conceptual Development of Quantum Mechanics* (New York: McGraw-Hill, 1966), pp. 28–29.

13 See *Myth and Muse*, pp. 118–20, and *The Conceptual Development of Quantum Mechanics*, pp. 325–26.

14 See *Myth and Muse*, pp. 117–22.

15 These assumptions about the new physics will be detailed more fully in chapter 3. For a catalogue of the uses and abuses of the new physics, especially in literature, see *Myth and Muse*, pp. 21–24, 71–109, 128–53. Donley and Friedman point out that Henri Bergson, the French philosopher who opposed both materialism and mechanistic thought, may actually have influenced some of the physicists (pp. 9 and 124). See also Boris Kuznetsov, "Einstein, science and culture," in *Einstein: A Centenary Volume*, ed. A. P. French (Cambridge: Harvard University Press, 1979), p. 178. *Myth and Muse* is not the only book that examines the response of fiction writers such as Nabokov, Coover, and Woolf, or of poets such as Frost and MacLeish to science. Although they are not, as I am, primarily interested in modernist responses to science in the American context, see also Hyatt Howe Waggoner, *The Heel of Elohim: Science and Values in Modern American Poetry* (Norman: University of Oklahoma Press, 1950); Ian F. A. Bell, *Critic As Scientist: The Modernist Poetics of Ezra Pound* (New York: Methuen, 1981), and passing

remarks in Hugh Kenner, *A Homemade World: The American Modernist Writers* (New York: Alfred A. Knopf, 1975); Joseph N. Riddel, *The Inverted Bell: Modernism and the Counterpoetics of William Carlos Williams* (Baton Rouge: Louisiana State University Press, 1974), and *Skyscraper Primitives*.

16 *Myth and Muse*, p. 152.

17 Arthur I. Miller, "Visualization Lost and Regained: The Genesis of the Quantum Theory in the Period 1913–1927," in *On Aesthetics in Science*, ed. J. Wechsler (Cambridge: Massachusetts Institute of Technology Press, 1978), p. 75, cited and discussed in *Myth and Muse*, p. 121. See also Gerald Holton, *The Scientific Imagination: Case Studies* (Cambridge: Cambridge University Press, 1978), pp. 10–11.

18 Wyndham Lewis, *Time and Western Man* (New York: Harcourt, Brace, 1928); Edmund Wilson, *Axel's Castle: A Study in the Imaginative Literature Of 1870 to 1930* (1931; reprint ed., New York: Charles Scribner's Sons, 1959); Alfred North Whitehead, *Science and the Modern World: Lowell Lectures, 1925* (New York: Macmillan, 1925). See also, for example, Allen Tate, "Post-Symbolism" (review of *Axel's Castle*), *Hound & Horn* 4(July–September 1931):619–24; R. P. Blackmur, "The Enemy" (review of *Time and Western Man*), *Hound & Horn* 1 (March 1928):270–73; Claude Bragdon, "New Concepts of Time and Space," *Dial* 68(February 1920):187–91, or Thomas Jewell Craven, "Art and Relativity," *Dial* 70(May 1921):535–39.

19 *Science and the Modern World*, especially chapters 1, 5, 9, and 13.

20 *Myth and Muse*, p. 194.

21 See, for example, the articles by Ida M. Tarbell, "Louis Pasteur," *McClures* 19(1902):143–51; Cleveland Moffett, "The Conquest Of Five Great Ills," *McClures* 21(1903):484–90; the anonymous review of René Vallery-Radot, *The Life of Pasteur*, *Nation* 74(March 1902):192–94, and Stephen Paget's biography, *Pasteur And After Pasteur* (London: Adam and Charles Black, 1914). One might also note the impact that the awareness of hygiene had on American lives and American landscapes. Williams's *Autobiography* talks about the installment of sewers during his childhood (A 280). In *Middletown: A Study in Contemporary American Culture* (New York: Harcourt, Brace, 1929), Robert S. and Helen Merrell Lynd discuss the increasingly utilitarian approach to health education in the public schools (p. 190), although they also say that by the 1920s middle-American towns were still debating the virtues of germ theory and pasteurization (pp. 450 and 454). Harvey Green, *The Light of the Home: An Intimate View of the Lives of Women in Victorian America* (New York: Pantheon Books, 1983), pp. 40 and 134, notes that advertisements in women's magazines frequently emphasized the medical profes-

sion's recommendations about cleanliness and occasionally mentioned Pasteur by name.

22 But see *Skyscraper Primitives*, pp. 143–64. Tashjian's chapter on Crane also makes any lengthy discussion of him unnecessary here, since his analysis places Crane in the tradition I am attempting to outline.

23 Hart Crane, *The Complete Poems and Selected Letters and Prose of Hart Crane*, ed. Brom Weber (Garden City, New York: Doubleday, 1966), p. 219.

24 "An Enquiry," *New Verse*, 11(October 1934):16.

25 "An Enquiry," p. 15.

CHAPTER 1

1 Solyman Brown, "Essay on American Poetry," cited in James E. Miller, Jr., *The American Quest for a Supreme Fiction: Whitman's Legacy in the Personal Epic* (Chicago: University of Chicago Press, 1979), p. 25. In *Toward a National Taste*, without disputing that there was anxiety about whether America could produce a national art and literature, Meredith Neil argues that in fact America had developed a distinctive national taste as early as 1815.

2 "The Intellectual Life of America," *New Princeton Review* 6(November 1888):317 and 320. Quoted in Kermit Vanderbilt, *Charles Eliot Norton: Apostle of Culture in a Democracy* (Cambridge: Harvard University Press, 1959), p. 207.

3 *The Incorporation of America*, p. 144.

4 *Middletown*, p. 249. The Lynds' study confirms many of the criticisms of American culture voiced in the 1910s and 1920s. Williams later connected Paterson, New Jersey, with Muncie in his exploration of American life, *Paterson* (*Pat* 10). See also Richard Wightman Fox, "Epitaph for Middletown," *The Culture Of Consumption: Critical Essays in American History, 1880–1980*, ed. Richard Wightman Fox and T. J. Jackson Lears (New York: Pantheon Books, 1983), pp. 101–41, on what informed Robert Lynd's study and on its impact.

5 *Middletown*, pp. 232–33, 249–50.

6 *Middletown*, pp. 238–39.

7 *A History of Modern Poetry*, p. 95.

8 Cited in *A History of Modern Poetry*, p. 90.

9 *An American Anthology, 1787–1900* (New York: Houghton, Mifflin, 1900), p. xxviii. Jackson Lears connects the role of Victorian women as guardians of culture with the transition to a market economy (*No Place of Grace*, pp. 15–16). See also Ann Douglas, *The Feminization of American Culture* (New York: Knopf, 1977) and *A Genteel Endeavor*, es-

pecially pp. 113–35 on Stedman. Perkins (*A History of Modern Poetry*, p. 95) suggests that the genteel tradition was itself an attempt to compensate for a felt lack of culture.

10 6 April 1909, letter to Edgar Williams, YALC ([Yale American Literature Collection]. I am also grateful to the Beinecke Rare Book and Manuscript Library, Yale University, and to the Williams Estate for use of these materials). Williams is arguing against this idea, even in 1909.

11 "The Art Of Poetry VI," *Paris Review*, No. 32(Summer–Fall 1964):133.

12 "Some Hints Toward The Enjoyment Of Modern Verse," YALC.

13 "Art," in *Civilization in the United States: An Inquiry By Thirty Americans*, ed. Harold E. Stearns (New York: Harcourt, Brace, 1922), p. 229.

14 Van Wyck Brooks, "Harvard and American Life," *Contemporary Review*, 12 December 1908, p. 618. I would like to thank Casey Blake for allowing me to read portions of his University of Rochester dissertation, "Young America: The Cultural Criticism of Randolph Bourne, Van Wyck Brooks, Waldo Frank, and Lewis Mumford," which informs the discussion of Brooks in this chapter. See also James Hoopes, *Van Wyck Brooks: In Search of American Culture* (Amherst: University of Massachusetts Press, 1977).

15 "Harvard and American Life," p. 618.

16 Van Wyck Brooks, *The Wine of the Puritans: A Study of Present-Day America* (1908; reprint ed., Folcroft, Pennsylvania: Folcroft Library Editions, 1974).

17 See Siegfried Giedion, *Mechanization Takes Command: A Contribution to Anonymous History* (New York: Oxford University Press, 1948), pp. 40–44, 93–101.

18 "The Literary Life," in *Civilization in the United States*, p. 179. Brooks's later views were influenced by his exposure to the Fabian Societies in England and to the Morrisite crafts movement, which was more radical in England than in the United States (see "Young America").

19 For a study of the rise of "Young Intellectuals" in America from 1870 to 1930, see Martin J. Sklar, "On the Proletarian Revolution and the End of Political-Economic Society," *Radical America* 3(May–June 1969): especially pp. 18–36. Their forum was the small magazines where the poets I discuss also published.

20 *The Incorporation of America*, pp. 65–69, 52–54.

21 See *Anti-Intellectualism in American Life*, pp. 25–26, 37. See also *Critic As Scientist*, pp. 8–9, on how, from the other side of the Atlantic, Pound's scientific analogies tried to emphasize the professional quality of poetry.

22 See *Anti-Intellectualism in American Life*, p. 151, on the class assump-

tions behind anti-intellectualism. *Skyscraper Primitives*, pp. 22–23, 40, also discusses the relationship between the critiques of American commercialism and antirationalism as well as anti-intellectualism.

23 Van Wyck Brooks, *America's Coming-Of-Age* (New York: B. W. Huebsch, 1915), p. 164.
24 "An Enquiry," p. 16.
25 Wallace Stevens, "An Enquiry," p. 15; 31 January 1921, letter to [Ferdinand] Reyher, WAS 1559, Huntington Library. I am grateful to the Huntington Library, San Marino, California, for use of these materials.
26 Horace Gregory, "William Carlos Williams and the 'Common Reader'," *Briarcliff Quarterly* 3(October 1946):186.
27 Ellen Williams, *Harriet Monroe and the Poetry Renaissance: The First Ten Years of "Poetry," 1912–1922* (Urbana: University of Illinois Press, 1977), pp. 20–21.
28 Michael True, "Modernism, *The Dial*, and the Way They Were," in *The Dial: Arts and Letters in the 1920's*, ed. Gaye L. Brown (Worcester, Massachusetts: Worcester Art Museum, 1981), p. 13.
29 1922 (?) letter to Stevens, WAS 1158, Huntington Library.
30 Henry McBride, "Modern Art," *Dial* 70(April 1921):481–82.
31 "The Art of Poetry VI," p. 134.
32 *Skyscraper Primitives*, p. 87, notes some of the implicit tensions within the calls for an indigenous American art.
33 "Literature" and "Poetry" in *America Now: An Inquiry Into Civilization in the United States By Thirty-Six Americans*, ed. Harold E. Stearns (New York: Charles Scribner's Sons, 1938), pp. 37 and 48.
34 *Mechanization Takes Command*, pp. 96 and 115.
35 John Dewey, "Current Tendencies in Education," *Dial* 62(April 1917):288 and "Education and Social Direction," *Dial* 64(April 1918): 333.
36 Malcolm Cowley, *Exile's Return: A Literary Odyssey of the 1920's* (1934; rev. ed. New York: Viking Press, 1951), p. 94.
37 "Current Tendencies in Education," pp. 288–89.
38 See Blake's chapter "The Politics of Cultural Renewal" in "Young America" for a critique of the use of the word *science* in Dewey as well as in the early writings of Bourne and Brooks; the latter two writers, Blake argues, ultimately embraced Dewey's view of knowledge as experience without retaining Dewey's emphasis on scientific technique.
39 John Walker III, "Symbols of Europe and America," *Hound & Horn* 1(June 1928):361.
40 See *Literary Criticism*, ed. Lionel Trilling (New York: Holt, Rinehart and Winston, 1970), p. 341, and *Modern Literary Criticism*, ed. Lawrence I. Lipking and A. Walton Litz (New York: Atheneum, 1972), pp.

123–34, and I. A. Richards, *Science and Poetry* (London: Kegan Paul, Trench, Trubner and Company, 1926).

41 "The Enemy," p. 273.

42 "The Enemy," p. 273.

43 Review of *The Life of Pasteur*, p. 194.

44 Lionel Trilling, "The Sense of the Past," *Influx: Essays on Literary Influence*, ed. Ronald Primeau (Port Washington, New York: Kennikat Press, 1977), p. 23.

45 Leo Stein, "American Optimism," *Seven Arts* 2(May 1917):82 and 88. See also Leo Stein, "Reality," *Dial* 83(September 1927):207, where he makes a similar point, saying it "is a commonplace that our moral culture has not kept pace with our science."

46 *Civilization in the United States*, p. 157.

47 *The Incorporation of America*, p. 63.

48 *No Place of Grace*, p. 31.

49 See *A Genteel Endeavor*, especially p. 184.

50 M. H. Abrams, *Natural Supernaturalism: Tradition and Revolution in Romantic Literature* (New York: W. W. Norton, 1971), p. 68. The classic study of the Romantic treatment of science remains idem, "Science and Poetry in Romantic Criticism," *The Mirror and the Lamp: Romantic Theory and the Critical Tradition* (1953; reprint ed., New York: W. W. Norton, 1958), pp. 298–335.

51 "A Defense of Poetry," *The Selected Poetry and Prose of Shelley*, ed. Harold Bloom (New York: Signet, 1966), p. 442.

52 See Mary Shelley, *Frankenstein* (1818; reprint ed., New York: Bantam Books, 1981), p. xxvi, on the "hideous progeny" of science.

53 Thomas Love Peacock, "The Four Ages Of Poetry," in *The Four Ages Of Poetry, Etc.*, ed. H. F. B. Brett-Smith (Oxford: Basil Blackwell, 1921), p. 17.

54 "The myth of the postmodernist breakthrough," *Tri-Quarterly* 26(1973):391.

55 "Marxism and English Romanticism: The Persistence of the Romantic Movement," *Romanticism Past and Present* 6(1982):35.

56 See, for example, W. C. Blum, "American Letter," *Dial* 70(April 1921):566, where Blum humorously charges Brooks with advocating the improvement of morals and hoping art will be the result.

57 Crispin's decision, Stevens strongly suggests, might spell the end of poetry, as, for Stevens, poetry requires more than local color (*CP* 27–46).

58 *I Wanted To Write a Poem*, reported and edited by Edith Heal (Boston: Beacon Press, 1958), p. 17. Williams's final understanding of how poetry and reality were related is in practice, however, closer to Stevens's than to

that of the critics for the *Seven Arts*, although he did not object as strongly as Stevens to the localists' program.

59 *The Complete Poems and Selected Letters and Prose of Hart Crane*, pp. 144–46. See also Alan Trachtenberg, *Brooklyn Bridge: Fact and Symbol* (1965; second ed., Chicago: University of Chicago Press, 1979), pp. 148–55, on how, throughout his work, Crane established a clear antagonism between the mythic consciousness or language of poetry and technological America, especially urban America.

60 *Mechanization Takes Command*, pp. 685–703 (on America and domestic labor saving devices and on modern bathrooms), 561–95 (on washing machines, dishwashers, vacuum cleaners and irons), 40–41 (on advertisements for automobiles and kitchens in America). See also *Skyscraper Primitives*, p. 7, on the definition of technology as American.

61 "Marginal Notes on *Civilization in the United States*," *Dial* 72(June 1922):555. For an excellent discussion of technology's place in early definitions of American culture and development, see John F. Kasson, *Civilizing the Machine: Technology and Republican Values in America, 1776–1900* (New York: Grossman Publishers, 1976).

62 "Education and Social Direction," p. 333, and *The Letters of Randolph Bourne: A Comprehensive Edition*, ed. Eric J. Sandeen (Troy, New York: Whitson Publishing Company, 1981), p. 263 [30 July 1914, letter to Alyse Gregory]. In this definition, efficiency is equated with modernity.

63 *The Complete Poems and Selected Letters and Prose of Hart Crane*, pp. 261–62, 232.

64 *America's Coming-Of-Age*, p. 44.

65 *Dial* 68(February 1920):240.

66 Dewey's understanding of localism, in fact, was more sophisticated than Oppenheim's. In "Americanism and Localism," *Dial* 68(June 1920): 686–87, Dewey explicitly argues that using "local colour" is a superficial way of including local reality in literature; he further suggests that generalized stories about America found in popular journals such as the *Saturday Evening Post* had no depth, although they may have stemmed from the mobility of Americans in the age of cars and railroads, which gave people superficial views of places through which they passed.

67 *America's Coming-Of-Age*, p. 112.

68 See the first chapters of Stephen Tapscott, *American Beauty: William Carlos Williams and the Modernist Whitman* (New York: Columbia University Press, 1984). Other views of Whitman will also be discussed in chapter 3.

69 *The Portable Walt Whitman* (1945; rev. ed., New York: Viking, 1974), p. 264. See also "I Hear America Singing" (p. 182).

70 Leo Marx, *The Machine in the Garden: Technology and the Pastoral Ideal in America* (New York: Oxford University Press, 1964), p. 222.
71 *The Portable Walt Whitman*, p. 140.
72 *The Machine in the Garden*, pp. 224–25.
73 *The Portable Walt Whitman*, pp. 278–79.
74 See *The Incorporation of America*, pp. 61–62. I will argue that Whitman's legacy was changed by events and circumstances of the early part of the century, however. See also Cary Nelson, *Our Last First Poets: Vision and History in Contemporary American Poetry* (Urbana: University of Illinois Press, 1981), on further changes affecting the Whitman tradition after 1960.
75 *America's Coming-Of-Age*, p. 44.
76 Marsden Hartley, "Modern Art in America," *Adventures in the Arts* (New York: Boni and Liveright, 1921), p. 60.
77 The January 1927 announcement by Scofield Thayer and J. S. Watson is cited in *Predilections*, p. 134.
78 "American Letter," p. 566.

CHAPTER 2

1 "Americanism and Localism," p. 687.
2 Waldo Frank, *Memoirs of Waldo Frank*, ed. Alan Trachtenberg (Amherst: University of Massachusetts Press, 1973), especially p. 63–64 and 83–95.
3 Kozloff, *Cubism/Futurism*, p. 5; John Adkins Richardson, *Modern Art and Scientific Thought* (Urbana: University of Illinois Press, 1971), p. 104; *Modern Art*, p. 140; *Cubism*, p. 180–81.
4 Gleizes and Metzinger, "Cubism," in *Modern Artists on Art: Ten Unabridged Essays*, ed. Robert L. Herbert (Englewood Cliffs: Prentice-Hall, 1964), p. 7.
5 For a discussion of the American interest in the pace of modern life, not necessarily related to the artists' statements about speed, see Cecelia Tichi, "Twentieth Century Limited: William Carlos Williams' Poetics of High-Speed America," *William Carlos Williams Review* 9(Fall 1983):49–72.
6 Wyndham Lewis, *The Soldier of Humour and Selected Writings* (New York: New American Library, 1966), p. 249.
7 *Cubism/Futurism*, p. 65.
8 *Cubism*, p. 206.
9 *Cubism*, p. 222.
10 *Modern Art and Scientific Thought*, p. 136.

11 *Cubism/Futurism*, pp. 99–100.

12 André Breton, "Le Manifeste du Surréalisme," in *Surrealism*, ed. Patrick Waldberg (New York: McGraw–Hill, 1965), p. 72, cited in *Modern Art and Scientific Thought*, p. 141.

13 *Surrealism*, p. 28; *Modern Art and Scientific Thought*, pp. 141–42.

14 *Modernism: The Critical Idiom*, pp. 19–20.

15 Allen Upward, "The New Age," *New Age*, 26 January 1911, p. 297.

16 "Cubism," p. 15.

17 *Dada and Surrealist Art*, p. 56.

18 *Modern Art*, p. 136; *Cubism*, p. 241–44.

19 Reprinted in *Léger and Purist Paris* (London: The Tate Gallery, 1970), p. 88.

20 Diana Collecott Surman, "Towards The Crystal: Art and Science in Williams' Poetic," in *William Carlos Williams: Man and Poet*, ed. Carroll F. Terrell (Orono, Maine: The National Poetry Foundation at the University of Maine, 1983), p. 187.

21 *Modern Art*, p. 169; *Cubism*, pp. 204, 222, 241–44; *Cubism/Futurism*, pp. 148–50.

22 *Cubism/Futurism*, pp. 110–11; *Cubism*, pp. 204 and 241–44.

23 *Cubism/Futurism*, p. 122. As early as 1911, Metzinger also related mental motion to Einsteinian physics, an association that will be discussed in chapter 3. See Jean Metzinger, "Cubisme et tradition," *Paris-Journal*, 18 August 1911, cited in *Modern Art and Scientific Thought*, pp. 109–10. It is worth pointing out that Einstein himself wrote that "this new artistic 'language' has nothing in common with the Theory of Relativity." His letter stating this is reprinted in Paul M. Laporte, "Cubism and Relativity," *Art Journal* 25(Spring 1966):246–48, and cited in *Modern Art and Scientific Thought* in a long refutation of there being an actual relationship between Cubism and relativity theory (pp. 105–12).

24 *Futurism*, p. 131.

25 *Mechanization Takes Command*, p. 27.

26 *Cubism*, pp. 156, 204, 222, 241–44.

27 *New Republic*, 3 August 1921, p. 264.

28 Lewis Mumford, "The City," in *Civilization in the United States*, p. 12. For an account of Mumford's shifts in attitude toward technology throughout his career, see Christopher Lasch, "Lewis Mumford and the Myth of the Machine," *Salmagundi*, no. 49 (Summer 1980):4–28.

29 Lewis Mumford, *American Taste* (San Francisco: Westgate Press, 1929), p. 25.

30 See the introduction to *Alfred Steiglitz: Camera Work*, ed. Marianne

Fulton Margolis (New York and Rochester: Dover Publications and the International Museum of Photography at George Eastman House, 1978), pp. vii–xi.

31 *Towards a New Architecture,* trans. Frederick Etchells (London: John Rodker Publisher, 1931), p. 31.

32 *Towards a New Architecture,* pp. 15, 95, 237.

33 *Towards a New Architecture,* p. 89.

34 See Henry Sayre's "After *The Descent of Winter:* Objectivism, Precisionism, and the Aesthetics of the Machine," pp. 18 and 29, in which Sayre discusses Williams and Le Corbusier, noting that Williams owned the French edition of Ozenfant and Le Corbusier's *La Peinture Moderne* (Paris: G. Cres, 1927), which he probably purchased in Paris in 1927. I am grateful to Professor Sayre for providing me with a copy of his paper, which was presented at the 1985 session on Williams at the Modern Language Association meetings in Chicago.

35 "For A New Magazine," *Blues* 1(March 1929):31.

36 "The Basis of Poetic Form," in "Notes for Talks and Readings," YALC (Copyright © 1983 by William Eric Williams and Paul H. Williams); "A Few General Correctives to the Present State of American Poetry," YALC.

37 "Comment," *Dial* 79(August 1925):177.

38 "Art," p. 241.

39 See Hitchcock's essays as well as the "Foreword" by Alfred H. Barr, Jr., pp. 14–15, and Philip Johnson's "Historical Note," p. 19, in *Modern Architecture.*

40 "The Decline of Architecture," *Hound & Horn,* 1(September 1927):34 and 30.

41 No. 13 (Summer 1928): 245.

42 *Mechanization Takes Command,* pp. 701–03.

43 This taste was first defined by the European art movements of Cubism, Futurism, and Dada. See *Skyscraper Primitives,* especially p. 7.

44 Williams, "Studiously Unprepared," YALC.

45 *Towards a New Architecture,* especially pp. 11 and 31.

46 *Towards a New Architecture,* pp. 95 and 107. In "After *The Descent of Winter,*" Henry Sayre suggests that Williams's 1944 definition may be a conscious echo of Le Corbusier's.

47 See Frank Doggett's suggestion in *Wallace Stevens: The Making of the Poem* (Baltimore: Johns Hopkins University Press, 1980), p. 98, that here Stevens owes a debt to Santayana who, in *Realms of Being,* called the flux of the natural world the machine of nature. Stevens's use of mechanistic images is rare, except during the 1930s (see chapter 6).

48 *Pagany* 3(Winter 1932):142.

49 24 November 1940 letter to James Laughlin, YALC. See "Towards the

Crystal" (p. 188), on how images of crystals were used by Le Corbusier and Ozenfant in their description of Purism.

50 In a paper presented at the 1985 session on Williams at the Modern Language Association meetings in Chicago, Cecelia Tichi argued that Williams's use of technological imagery was influenced by the prevalence of such images both in popular culture and in medical texts ("Medicine and Machines Made of Words.") I am grateful to Cecelia Tichi for providing me with a copy of this paper. See also *The Incorporation of America* for a discussion of how, in the 1880s, images "of machinery filtered into the [American] language, increasingly providing convenient . . . metaphors for society and individuals" (pp. 44–47).

51 "American Letter," p. 565. In "Twentieth Century Limited," p. 69, Cecelia Tichi also notes that Williams would have known Osler's work.

52 *Towards a New Architecture*, p. 8, see also pp. 89, 232, 237, 275.

53 Malcolm Bradbury and James McFarlane, "The Name and Nature of Modernism," *Modernism*, p. 33.

54 T. E. Hulme, "Modern Art And Its Philosophy," *Speculations*, ed. Herbert Read (New York: Harcourt, Brace, 1924), p. 96.

55 "Modern Art And Its Philosophy," pp. 108–09.

56 T. E. Hulme, "Searchers After Reality," *Further Speculations*, ed. Sam Hynes (Lincoln: University of Nebraska Press, 1962), p. 17. The essay was first published in the *New Age* in 1909.

57 For examples of Georgian poetry, see E[dwin Howard] Marsh, *Georgian Poetry, 1911–1912* (London: Poetry Bookshop, 1912). Marsh's "Prefatory Note to First Edition" suggests that even Georgian poets, whose work was not modernist, felt a literary revolution was in the air, although most of the contributors to Marsh's volume did not follow Pound or the artists.

58 Ezra Pound, *The Selected Letters of Ezra Pound, 1907–1941*, ed. D. D. Paige (1950; reprint ed., New York: New Directions, 1971), p. 6. In spite of Pound's rhetoric, the collection *A Lume Spento* is most strongly reminiscent of Pre-Raphaelite poetry. See Thomas H. Jackson, *The Early Poetry of Ezra Pound* (Cambridge, Massachusetts: Harvard University Press, 1968), pp. 23, 29, 157, 172–73.

59 *A Lume Spento and Other Early Poems* (New York: New Directions, 1965), p. 7.

60 "A Few Dont's," *Poetry*, 1 (March 1913), reprinted in *The Literary Essays of Ezra Pound* (Norfolk, Connecticut: New Directions, 1954), pp. 4–5.

61 *Literary Essays*, p. 46. For a full description of Pound's uses of science and of his rhetoric, see *Critic As Scientist* and John T. Gage, *In The Arresting Eye: The Rhetoric of Imagism* (Baton Rouge: Louisiana State University Press, 1981).

62 See, for example, Blum's "American Letter," p. 565, or Laura Riding and Robert Graves, *A Survey of Modernist Poetry* (1927; reprint ed. New York: Folcraft Library Editions, 1971), especially pp. 216–17 on Stevens, who did not consider himself an Imagist.
63 See especially Henry M. Sayre, *The Visual Text of William Carlos Williams* (Urbana: University of Illinois Press, 1983).
64 *The Visual Text of William Carlos Williams*, p. 69, and Bram Dijkstra, *The Hieroglyphics of a New Speech: Cubism, Stieglitz, and the Early Poetry of William Carlos Williams* (Princeton: Princeton University Press, 1969), p. 191.
65 *Modern Art*, pp. 154, 169, 174–75.
66 *The Visual Text of William Carlos Williams*, p. 61.
67 *The Visual Text of William Carlos Williams*, pp. 60–61. On the ambivalence involved in visual images of technology, see Donald B. Kuspit, "Individual and Mass Identity in Urban Art: The New York Case," *Art in America* 65(September–October 1977):67–77. Dickran Tashjian, *Skyscraper Primitives*, p. 157, discusses Crane's tendency to humanize and pastoralize images of industrial machinery or architecture.
68 *The Visual Text of William Carlos Williams*, p. 63. It should be said that Sayre has in mind the tension between order and multiplicity, and between fact or matter and mind. The latter dyad, I will argue, was articulated in terms of the tension between a machine aesthetic and a more fluid style in the writings of Moore and Stevens as well as in Williams's work.
69 *The Selected Letters of Ezra Pound*, p. 10.
70 *The Selected Letters of Ezra Pound*, p. 9; the letter, dated August 18, 1912, is also to Harriet Monroe.
71 27 August 1925 letter to Stieglitz, YALC.
72 No. 13 (Summer 1928): 252.
73 See also *Critic As Scientist*, pp. 225–30, where Ian Bell argues that in England as well the poets' appeals to science, although a reaction against industrial capitalism, resulted in art being viewed as a commodity.
74 Kenneth Burke, *Counter-Statement* (Los Altos, California: Hermes Publications, 1931), p. 90.
75 "Comment," *Dial* 81(September 1926):268.
76 "Comment," *Dial* 85(December 1928):541; "Comment," *Dial* 81(October 1926):358.
77 *Middletown*, p. 237; Moore's experience as a librarian may have made her aware of such trends before the Lynds' study appeared.
78 *Middletown*, p. 237; reading on the fine arts increased only twenty-eight-fold, while the reading of literary fiction increased less than four-fold, according to the Lynds.

79 For an example of how hygiene was considered a major American achievement, see George Santayana's *The Last Puritan: A Memoir In the Form of a Novel* (New York: Charles Scribner's Sons, 1937), p 54, where the narrator comments after one of his American characters visits the Orient: "we had nothing good to teach the East, except indeed hygiene."
80 "American Optimism," p. 88.
81 *Counter-Statement*, p. 121.
82 "The Function of Literature, An Address to Be Given Before The Institute of Humanistic Studies for Executives of the University of Pennsylvania Program for Junior Executives of the Bell Telephone Companies," YALC.
83 "The Function of Literature."
84 *The Complete Poems and Selected Letters and Prose of Hart Crane*, p. 225.
85 *Civilization in The United States*, pp. 157 and 419–20.
86 "The Aesthetics of the Machine and Mechanical Introspection in Art," *Broom* 3(October 1922):236.
87 *Brooklyn Bridge*, p. 97.
88 Cited in Andrew Bergman, *We're In The Money: Depression America and Its Films* (New York: Harper Colophon, 1971), p. 94.
89 "American Optimism," p. 88. Notes on Stein's second article, "Tradition and Art," from *Arts* (May 1925) appear in Moore's Reading Diary for 1924–1930, Rosenbach 1250 / 5, MS, p. 81. I am grateful to the Marianne Moore Collection at the Rosenbach Museum and Library, Philadelphia, Pennsylvania, and to the Estate of Marianne C. Moore for use of these materials.
90 "Comment," *Dial* 83(October 1927):359.

CHAPTER 3

1 *The Complete Poems and Selected Letters and Prose of Hart Crane*, pp. 88–89. See John Carlos Rowe, "The 'Super-Historical' Sense of Hart Crane's *The Bridge*," *Genre* 11(Winter 1978):601, on the destructiveness of modern technology in Crane's writings.
2 *Skyscraper Primitives*, p. 150. See also *A Genteel Endeavor*, pp. 124–25, on Stedman's attempt to equate poetic and scientific—in this case, Newtonian—force, and Ronald E. Martin's *American Literature and the Universe of Force* (Durham, North Carolina: Duke University Press, 1981), especially pp. xi, xiv, 59–60, 94–95, on the late nineteenth-century American equation of force, energy, and a deterministic, mechanical reality.
3 Charles Demuth, *The Blind Man*, no. 2 (May 1917):6. See also Henry

M. Sayre, "The Satyrs' Tribute to the Painters: The Tyranny of the Image: The Aesthetic Background," *William Carlos Williams Review* 9(Fall 1983):125–34, on the emphasis on act or process in American painting and poetry. As Sayre points out, the visual arts also appealed, on the one hand, to "object-free abstract action" and, on the other hand, to "objective representation" or a "return to the object" (p. 129).

4 *Time and Western Man*, pp. 152 and 163. See Sanford Schwartz, *The Matrix of Modernism: Pound, Eliot, and Early Twentieth-Century Thought* (Princeton: Princeton University Press, 1985), especially chapters 1 and 2, on the similarities between Bergsonian philosophy and other philosophies of the period.

5 *Hound & Horn* 1(December 1927):179.

6 *Time and Western Man*, p. 208; Lewis is quoting from Bertrand Russell's *The Philosophy of Bergson*.

7 Walter Lowenfels, *transition*, no. 14 (Fall 1928):105.

8 The section titles are reprinted in Emily Mitchell Wallace, A *Bibliography of William Carlos Williams* (Middletown, Connecticut: Wesleyan University Press, 1968), p. xix.

9 William Carlos Williams, "Vs," Buffalo, pp. 6(b) and 4(a). The essay was given as a talk at New York University and then published later the same month in *Touchstone* (New York) 1(January 1948):2–7 (A *Bibliography*, p. 214). I am grateful to the Poetry/Rare Books Collection, University Libraries, State University of New York at Buffalo for the use of these materials.

10 William E. Leverette, Jr., "E. L. Youman's Crusade for Scientific Autonomy and Respectability," *American Quarterly* 17(Spring 1965):12–32, and Robert H. Lowie, "Science," in *Civilization in the United States*, pp. 151–61.

11 *Anti-Intellectualism in American Life*, pp. 25–26.

12 See *Critic as Scientist*, pp. 1–79, 108–09, on Pound's responses to empirical science and to nineteenth-century *biological idealism*. Bell also suggests that whatever references to these sciences are found in early modernism are borrowed from Pound. See chapter 6 for a more complete discussion of the competing models of nineteenth-century science.

13 See *No Place of Grace*, p. 31, and *The Incorporation of America*, p. 63.

14 *Time and Western Man*, p. 141. See also *Myth and Muse*, pp. 17–20, on Einstein's popularity.

15 "Carpentier: A Symbol," *New Republic*, 20 July 1921, p. 206. See Carol C. Donley's account of popular press coverage of Einstein in " 'A little touch of / Einstein in the night—': Williams' Early Exposure to the Theories of Relativity," *William Carlos Williams Newsletter* 4(Spring 1978):10–13.

16 *Contact* 1 (January 1921). Charles Proteus Steinmetz in *Four Lectures on Relativity and Space* (1923; reprint ed. New York: Dover, 1967), p. 23, which Williams read later, noted that "energy is the only real existing entity," while Whitehead, in *Science and the Modern World*, p. 102, proclaimed the "reality is the process."

17 "Einstein and The Poets," *Broom* 1(November 1921):84–88, includes subtitles such as the "Weight of Light, Deflection of Solar Rays, Non-Euclidean Warps in Space, Anti-Newtonian Substitutes for Gravitation and Time as a Fourth Dimension" (p. 84).

18 *The Letters of Hart Crane, 1916–1932* ed. Brom Weber (Berkeley and Los Angeles: University of California Press, 1965), p. 311.

19 "Periodical Reviews," *Hound & Horn* 1(September 1927):72; "The Enemy," pp. 270–73.

20 Cited, *Skyscraper Primitives*, p. 17.

21 See *The Scientific Imagination*, pp. 236–40, on the criteria used to judge good scientific work, and the way in which the scientific establishment, perhaps misleadingly, has fostered the image of science as logical and impersonal, not primarily as creative.

22 *Complete Poems and Selected Letters and Prose of Hart Crane*, p. 228.

23 *Complete Poems and Selected Letters and Prose of Hart Crane*, p. 225.

24 "The Turn of the view toward poetic technique," YALC.

25 Elaine Barry, *Robert Frost on Writing* (New Brunswick, New Jersey: Rutgers University Press, 1973), p. 118.

26 "The Decline of Architecture," p. 28.

27 De Zayas, "Modern Art: Theories and Representation," *Camera Work*, no. 44 (October 1913):14; cited in *Skyscraper Primitives*, p. 28; see also p. 32.

28 Quoted from Picabia in "Picabia, Art Rebel, Here to Teach New Movement," *New York Times*, 16 February 1913, section V, p. 9. This article does not mention Einstein, and it is cited as an example of the stereotypes of America that were later associated with Einstein's universe.

29 *Time and Western Man*, p. 437; it might be added that unlike Moore, Lewis was unhappy about this picture.

30 Notes for a talk at Brandeis, YALC.

31 See Perry Hobbs, "A Problem in Common Sense," *Hound & Horn* 1(December 1927):153–56, reviewing Russell's *The Analysis of Matter*, for a popular expression of this view. What might be called the *desolidification* of the world began with breakthroughs in nineteenth-century science, when energy had already begun to shift its vocabulary. The idea did not capture public attention, however, until the popularity and misapprehension of Einstein brought theoretical physics into the popular arena. For discussions of Maxwell and other figures in nineteenth-cen-

tury science, see *Critic as Scientist*, p. 136, and James R. Newman, *Science and Sensibility* (New York: Simon and Schuster, 1961), especially "James Clerk Maxwell," pp. 139–93.

32 The quote is from *Axel's Castle*, p. 158, which Williams read in the late 1940s (Paul Mariani, *William Carlos Williams: A New World Naked* [New York: McGraw-Hill, 1981], p. 557).

33 *No Place of Grace*, pp. 10–11; see also "Twentieth Century Limited," pp. 50–56, on the associations between speed and America.

34 Stuart Chase, "Men and Machines: A Billion Horses," *New Republic*, 6 March 1929, p. 61. See also Williams's 1923 *The Great American Novel*, which presents a playful examination of progress in America; the *heroine* of Williams's book is a Ford motor car.

35 Max Rychner, "Inquiry Among European Writers Into the Spirit of America," *transition*, no. 13 (Summer 1928): 258–59.

36 *No Place of Grace*, pp. 57 and 142. See also Michael Colacurcio, "The Dynamo and the Angelic Doctor: The Bias of Henry Adams' Medievalism," *American Quarterly* 17(Winter 1965):696–712, and *Skyscraper Primitives*, p. 18, on the particularly American association early in the century between primitivism and technology.

37 Kenneth Rexroth, "A Letter to William Carlos Williams," *Briarcliff Quarterly* 3(October 1946):193.

38 *Exile's Return*, pp. 19–20; Gissing's dislike of science was often cited; see, for example, Robert Shafer, "The Definition of Humanism," *Hound & Horn* 3 (July–September 1930):553. George Santayana, in "Reason In Science," *The Life of Reason or The Phases of Human Progress* (1905; reprint ed. New York: Charles Scribner's Sons, 1954), p. 385, notes that many saw science as barring "sympathy with the reality."

39 *Time and Western Man*, p. 416.

40 *Science and the Modern World*, p. 196.

41 William Carlos Williams, "Reply to a Young Scientist Berating Me Because of My Devotion to a Matter of Words," *Direction* 1(Autumn 1934):28.

42 *Science and Sensibility*, p. 291; *The Conceptual Development of Quantum Physics*, p. 329; *Four Lectures on Relativity and Space*, p. 23. See also *Myth and Muse*, pp. 9, 122–28.

43 *Science and Sensibility*, p. 482; *Einstein: A Centenary Volume*, p. 148.

44 *Physics and Philosophy: The Revolution in Modern Science* (New York: Harper and Brothers, 1958), p. 55.

45 Niels Bohr, *Atomic Theory and The Description of Nature* (Cambridge: Cambridge University Press, 1934), pp. 116–17.

46 J. T. Fraser, "Comments—Of Time and Proper Time," *The Voices of Time* (Amherst: University of Massachusetts Press, 1981), p. 478; *Science and Sensibility*, pp. 489 and 498–500. See *Myth and Muse*, p. 9.

47 Bertrand Russell, *Let the People Think: A Selection of Essays*, Thinker's Library, no. 84 (London: Watts, 1941), p. 48.
48 *Physics and Philosophy*, p. 144.
49 *Physics and Philosophy*, p. 179.
50 William Carlos Williams, "What Is the Use of Poetry?" (circa 1926) Buffalo, MS, pp. 6 and 10.
51 A 4 July 1931 *Illustrated London News* article on Anton Reiser's biography of Einstein is noted in Moore's Reading Diary for 1930–1943, Rosenbach 1250/6, MS, p. 29; the same notebook (pp. 162–72) mentions articles from *Scientific Monthly*. Examples of references to Russell (and other books on science) can be found in the Reading Diary for 1916–1921, Rosenbach 1250/2, MS, pp. 86 and 135.
52 Wallace Stevens, "Sur Plusieurs Beaux Sujects" [sic], II, Huntington, MS, pp. 1–2. Stevens's 1951 "A Collect of Philosophy" also mentions Alexander (*OP* 193–94) as does Lewis throughout *Time and Western Man* (see pp. xi–xiii, 443–56 and passim).
53 Samuel Alexander, *Space, Time, and Deity*, Volume I (New York: Macmillan, 1920), p. 48. For a good summary of Alexander's work, see A. Cornelius Benjamin, "Ideas of Time in the History of Philosophy," in *The Voices of Time*, pp. 25–29.
54 *A New World Naked*, p. 492; Lee Schultz, "The Doctor-Poet of Paterson and the Science of Art," Ph.D. diss., University of Tulsa 1977, p. 199, suggests that Williams read Eve Curie's *Madame Curie*.
55 "'A little touch of / Einstein in the night—'," pp. 10–13.
56 Mike Weaver, *William Carlos Williams: The American Background* (Cambridge: Cambridge University Press, 1971), pp. 46–49.
57 C. E. M. Joad, "Another Great Victorian," *New Statesman*, 2 March 1940, pp. 280 and 282, and C. E. M. Joad, "Henri Bergson," *New Statesman*, 11 January 1941, p. 34, cited in "The Noble Rider and The Sound of Words" (NA 25). In light of the poets' use of popular sources, most of the references to Einsteinian physics in this chapter are taken from books or articles that were read by Williams, Stevens, or Moore.
58 "The Enemy," pp. 272 and 271.
59 *Complete Poems and Selected Letters and Prose of Hart Crane*, p. 239.
60 *Science and the Modern World*, p. 167.
61 "The Enemy," p. 272.
62 *Let the People Think*, pp. 49–50. The book was found in Stevens's library, and is now housed at the Huntington.
63 Morris R. Cohen, "Roads to Einstein," *New Republic*, 6 July 1921, p. 174.
64 For the painters' use of Einstein, see Barbara Novak, *American Painting of the Nineteenth Century: Realism, Idealism, and the American Experience* (New York: Praeger, 1969), p. 262.

65 *Science and the Modern World*, pp. 282–83; the chapter, "Requisites for Social Progress", is one Williams found particularly exciting (*SL* 79 and 84–85).

66 See, for instance, *Time and Western Man*, pp. 347–48.

67 *Time and Western Man*, "An Analysis of the Mind of James Joyce," p. 86.

68 *The Visual Text of William Carlos Williams*, p. 74.

69 James Joyce, *Ulysses* (New York: Vintage Books, 1961), p. 37.

70 *Time and Western Man*, pp. x and 337, for example. *Myth and Muse*, pp. 102–06, compares Joyce's work with Einstein's universe and notes how often Joyce's style is said to be related to twentieth-century science. Donley and Friedman argue that *Ulysses* cannot be shown to refer specifically to Einsteinian physics and that there is no scientific style, unless one looks to the style conventionally used in writing about science. My point, however, is not that Joyce actually used Einstein's findings or that science dictates one style of writing rather than another, but that the style of *Ulysses* was perceived as being the style dictated by Einstein's discoveries. *Myth and Muse* does point out the association between the new physics and motion (p. 26).

71 *American Beauty* provides an excellent account of how American modernist poets reinvented Whitman in their need for an American tradition that linked private and public experience (as Einstein linked poetry to the world). For Stevens's use of Whitman, see Diane Wood Middlebrook, *Walt Whitman and Wallace Stevens* (Ithaca: Cornell University Press, 1974).

72 Stanley Burnshaw, "Turmoil in the Middle Ground," *New Masses*, 1 October 1935, p. 42.

73 Moore's admiration of Emerson is noted in Laurence Stapleton, *Marianne Moore: The Poet's Advance* (Princeton: Princeton University Press, 1978), p. 58, and it was recognized quite early by Alfred Kreymborg, in *A History of American Poetry: Our Singing Strength* (1929; reprint ed. New York: Tudor, 1934), p. 74.

74 *The Visual Text of William Carlos Williams*, pp. 73–74.

75 Williams was also at times capable of linking this view of poetry with scientific methods of working—as when he compared himself to Herbert Clark, patiently "watching . . . the field," sifting through facts (cited in *A New World Naked*, pp. 732–33)—just as he was able at times to describe an action aesthetic in terms of technology (see "Twentieth Century Limited," pp. 65–66).

76 *Time and Western Man*, p. 113.

77 *Time and Western Man*, p. 151, citing *Space, Time, and Deity*.

78 Quoted from Williams's first draft of the entry, which is in the Yale

Collection. Zukofsky's and others' understandings of objectivism differed from Williams's, although they too appealed to both models and to analogies with technology and science. See David McAleavey's "If to Know is Noble: The Poetry of George Oppen," Ph.D. diss., Cornell University 1975, especially the introduction and first chapter.

79 The references to Einstein are proposed in *The Heel of Elohim*, pp. 171–84. The passage from "Cape Hatteras" is in *The Complete Poems and Selected Letters and Prose of Hart Crane*, p. 89.

80 "A Problem In Common Sense," p. 156.

81 William Carlos Williams, "Let us order our world," *William Carlos Williams Review* 8(Fall 1982):18.

82 Robert McAlmon, "Romance X," *The Portrait of a Generation*, (Paris: Three Mountains Press, 1926), p. 39. Stevens's copy is to be found in the Huntington Library's collection of the books he owned.

83 See, for instance, Bertrand Russell, *The ABC of Relativity* (1925; rev. ed., rpt. New York: Mentor Books, 1959), p. 29, or *Four Lectures on Relativity and Space*, pp. 24–37.

84 "The Row Among the Physicists," *Nation*, 27 December 1919, p. 819.

85 It is relevant that Williams's most defensive comments are often found in his letters to Burke, whose own rational bent tended to push Williams toward defending poetry in opposition to both science and philosophy.

86 *Complete Poems and Selected Letters and Prose of Hart Crane*, p. 262.

87 2 November 1955 letter to Young, YALC.

88 "A Cry in the Night. Definition," Notes for Talks and Readings, YALC.

89 "Post-Symbolism," p. 622.

90 19 January 1956 letter to Carolyn Brown, YALC.

91 *Axel's Castle*, p. 298; *The ABC of Relativity*, p. 9.

92 See, for instance, Kenneth Burke, *Grammar of Motives* (New York: Prentice-Hall, Inc., 1945), pp. 485–502, especially 486, and L. S. Dembo, *Conceptions of Reality in Modern American Poetry* (Berkeley and Los Angeles: University of California Press, 1966), pp. 109–12.

CHAPTER 4

1 Discussions of Williams's fascination with modern art are numerous. For the most extended consideration of the topic, see *The Hieroglyphics of a New Speech; Skyscraper Primitives;* Dickran Tashjian, *William Carlos Williams and the American Scene, 1920–1940* (New York: Whitney Museum of American Art, 1978); William Marling, *William Carlos Williams and the Painters, 1909–1923* (Athens: Ohio University Press, 1982); and Williams's own statements in *A Recognizable Image: William Carlos Williams on Art and Artists*, ed. Bram Dijkstra (New York: New

Directions, 1978). *Skyscraper Primitives, William Carlos Williams: The American Background*, and A *New World Naked*, provide the most comprehensive notes on Williams, and Carol C. Donley's article, "Relativity and Radioactivity in William Carlos Williams' *Paterson*," *William Carlos Williams Newsletter* 5(Spring 1979):6–11, cites many of the papers, articles, and theses which show the rising interest in Williams's relationship to science.

2 See, for example, "Towards the Crystal" or also in *Man and Poet*, Marjorie Perloff, " 'To Give a Design': Williams and the Visualization of Poetry," pp. 159–86. The 1985 session on Williams at the Modern Language Association meetings in Chicago also included papers on this topic by Cecelia Tichi ("Medicine and Machines Made of Words") and Henry M. Sayre ("After *The Descent of Winter*: Objectivism, Precisionism, and the Aesthetics of the Machine").

3 "American Optimism," p. 79.

4 "American Optimism," pp. 82–84, 88.

5 Marcel Duchamp, "The Ricahrd Mutt Case," *The Blind Man*, no. 2(May 1917):5.

6 See, for instance, *The Wine of the Puritans*, pp. 14–15.

7 Williams's association with the *Seven Arts* is well documented: see, for instance, *The Hieroglyphics of a New Speech*, pp. 42 and 108. Williams himself mentions both the *Seven Arts* (A 147) and *The Blind Man* (*I* 10), as well as commenting specifically on Duchamp's "Fountain" (A 134).

8 *Adventures in the Arts*, p. 61; Henry M. Sayre, "Avant-Garde Dispositions: Placing *Spring and All* in Context," *William Carlos Williams Review* 10(Fall 1984):13–17.

9 See "The Satyrs' Tribute to the Painters," p. 125, discussing Williams's anecdote about Hartpence (A 240), and "Avant-Garde Dispositions," pp. 20–22, about Demuth's pun on materiality in his 1921 or 1922 painting, *Spring*, which Williams probably knew. See also James E. Breslin, "William Carlos Williams and Charles Demuth: Cross-Fertilization in the Arts," *Journal of Modern Literature* 6(April 1977):248–63.

10 Williams's brother was an architect who would have introduced Williams to the work of Le Corbusier and other modern architects, as witnessed in Williams's later essay, "The Basis of Faith in Art." Critics often refer Williams's stylistic experiments to the work and theories of Pound, in his imagist phase, or to the painters of the Arensberg circle. See, for example, Henry M. Sayre, "Distancing 'The Rose' from *Roses*," *William Carlos Williams Newsletter* 5(Spring 1979):18–19; "After *The Descent of Winter*"; "Towards the Crystal"; " 'To Give a Design' "; "Cross-Fertilization in the Arts"; *The Hieroglyphics of a New Speech*; *Williams: The American Background*, p. 29; or Williams's own statement in the *Autobiography* (A 148).

11 See Rod Townley, *The Early Poetry of William Carlos Williams* (Ithaca: Cornell University Press, 1975), p. 68, on Williams's "allegiance to the cult of experience." As late as 1962, despite his many accomplishments and despite the recognition he and his work had already begun to receive, Williams still revealed his mixed feelings about whether ordinary Americans could value poetry, and whether Rutherford had been the best place for him to write poetry. In "The Art Of Poetry VI," Stanley Koehler reports saying to Williams: "Eliot went to England; you stayed here." Williams responded, "To my sorrow" (p. 124). Although Williams then qualified his answer, his qualms about America's relationship to poets and poetry show throughout the interview.

12 *Complete Poems and Selected Letters and Prose of Hart Crane*, p. 219; Crane's "Modern Poetry" repeatedly returns to the question of poetry's function or use (see pp. 261–62, especially). See *Skyscraper Primitives*, p. 147, and *Brooklyn Bridge*, pp. 146–65, 168, for discussions of how Crane's poems do not accommodate or even in some cases establish a dialogue or dialectic with, history. On the other hand, Alan Trachtenberg's "Cultural Revisions in the Twenties: Brooklyn Bridge as 'Usable Past'," in *The American Self: Myth, Ideology, and Popular Culture*, ed. Sam B. Girgus (Albuquerque: University of New Mexico, 1981), pp. 58–75, offers a convincing analysis of Crane's relationship to American culture.

13 Arturo Schwarz, "An Interview with Man Ray: 'This is Not for America'," *Arts Magazine* 51(May 1977):117.

14 *The Wine of the Puritans*, p. 32. See also Lewis Mumford, "Lyric Wisdom," in *Paul Rosenfeld: Voyager in the Arts*, ed. Jerome Mellquist and Lucie Wiese (New York: Creative Age Press, 1948), pp. 43–45. Williams, as well as Moore and Stevens, contributed essays to this volume.

15 For a more careful study, see Henry F. May, *The End of American Innocence* (1959; reprint ed., Chicago: Quadrangle Books, 1964) on practical idealism. See also "Lewis Mumford and The Myth of the Machine," p. 4, for a note linking the optimism of some writers with the revolt against positivism.

16 Pound had a response. He wrote to Williams: "The thing that saves your work is *opacity* . . . NOT an American quality" (*The Selected Letters of Ezra Pound*, p. 124). Williams reprinted the letter in *Kora in Hell* (*I* 11). See also *Skyscraper Primitives*, pp. 87ff., for a brief but insightful discussion of this passage as relating Williams to the theories of the European modernists, the Dadaists in particular, even as Williams attempted to define an American context for art. As Tashjian remarks, Williams's "reasoning generated certain tensions" (p. 87). Williams's definition of inventive intelligence, which he insists on appropriating for America,

owes a debt not only to Dadaism, but to ideas being formulated in a number of other artistic circles. Stravinsky, for instance, distinguishes between imagination or fantasy and "invention [which] is not conceivable apart from actual working-out" (*Poetics of Music In The Form Of Six Lessons* [1947; reprint ed., New York: Vintage Books, 1956], p. 55).

17 In "Medicine and Machines Made of Words," Tichi argues that Williams's attraction to engineering in particular stemmed from his medical training, in which doctors were compared to engineers working on the human machine. I disagree with her only in my suggestion that Williams came to be self-conscious about some of the problems the use of technological metaphors posed.

18 Frank A. Manny, "John Dewey," *Seven Arts* 2(June 1917):223.

19 See Cecelia Tichi, "William Carlos Williams and the Efficient Moment," *Prospects: An Annual of American Cultural Studies*, ed. Jack Salzman, vol. 7 (New York: Burt Franklin, 1982), pp. 267–79, for a parallel discussion of the costs and benefits of Williams's acknowledgement of Taylorism in forging a poetics of efficiency, thus appropriating for poetry a movement that swept American businesses from 1912, and even affected American housewives. See also A *New World Naked*, pp. 86–87, 126, 176, on Williams's discomfort with the European scene and his subsequent increased, if at times uneasy, valorization of the American scene.

20 Here I will not fully discuss the way in which Williams takes the urban, industrial landscape as his main subject, but one should see James E. Breslin, *William Carlos Williams: An American Artist* (New York: Oxford University Press, 1970), pp. 66–68, for a suggestion on how, in poems like "At the Faucet of June," Williams criticizes industrialists, even while taking from them an ideal of beauty. See also poems like "View of A Lake" (*CEP* 96), *In the American Grain*, pp. 68 and 128, and the discussion of "Classic Landscape" in chapter 2.

21 See *The Visual Text of William Carlos Williams*, pp. 6, 80–81, and Stephen Cushman, *William Carlos Williams and the Meanings of Measure* (New Haven: Yale University Press, 1985) throughout.

22 " 'To Give a Design'," pp. 174–78; see also note 10, above.

23 *The Visual Text of William Carlos Williams*, p. 116, n. 7 and "Medicine and Machines Made of Words."

24 Bertrand Russell, "Is Nationalism Moribund?" *Seven Arts* 2(October 1917):682, and James Oppenheim, "Art, Religion and Science," *Seven Arts* 2(June 1917):229. See also Alan Trachtenberg, "Mumford in the Twenties: The Historian as Artist," *Salmagundi*, no. 49 (Summer 1980):33.

25 "Let us order our world," p. 18.

26 See "Towards the Crystal," pp. 187–88.

27 "After *The Descent of Winter,*" p. 18. Le Corbusier's definition is found in *Towards a New Architecture*, p. 95; see also pp. 4 and 107. It might be added that Le Corbusier also defines houses as poems (p. 263).

28 *Towards a New Architecture*, p. 203.

29 Emily Mitchell Wallace, "The Satyrs' Abstract and Brief Chronicle of Our Time," *William Carlos Williams Review* 9(Fall 1983):137.

30 *Contact*, no. 4 (Summar 1921):2.

31 See Ron Loewinsohn's "Introduction" to *The Embodiment of Knowledge*, pp. ix–xxv.

32 See Williams. *The American Background*, pp. 32–35, and *A New World Naked*, pp. 166 and 284, for a discussion of the exchanges between Robert McAlmon and Dewey, and between Williams and Dewey, on the issue of American education. I discuss *Spring and All* more fully in "Once More With Feeling: Teaching *Spring and All,*" *William Carlos Williams Review* 10(Fall 1984):7–12, where some of this discussion of Williams and Dewey first appeared.

33 "Education and Social Direction," p. 334.

34 "Current Tendencies in Education," p. 289.

35 "Education and Social Direction," p. 334.

36 John Dewey, *Democracy and Education: An Introduction to the Philosophy of Education* (1916; reprint ed., New York: Macmillan, 1920), p. 173.

37 "Medicine and Machines Made of Words."

38 *The Hieroglyphics of a New Speech*, pp. 190–91.

39 "Introduction," *William Carlos Williams: A Collection of Critical Essays*, ed. J. Hillis Miller (Englewood Cliffs, New Jersey: Prentice–Hall, 1966), p. 12.

40 In *Williams Carlos Williams: A Collection of Critical Essays*, pp. 3 and 8, Miller describes the tree as nonsymbolic, but he goes on to note that "Young Sycamore" exemplifies Williams's most characteristic mode of describing motion. See also Richard A. Macksey, "'A Certainty of Music': Williams' Changes," in *William Carlos Williams: A Collection of Critical Essays*, pp. 132–47, and J. Hillis Miller, *Poets of Reality: Six Twentieth-Century Writers* (New York: Atheneum, 1969), pp. 328–55, on Williams's *cycles*.

41 Indeed, the trunk's "undulant / thrust" is recalled by Williams's description of the undulant character of both poems and machines in the 1944 introduction to *The Wedge* (*SE* 256).

42 *Critic As Scientist*, pp. 244–46, contains an insightful critique of this side of Williams's poetics, as does *The Meanings of Measure*, p. 137, which proposes that Williams may draw on Emerson for such ideas.

43 A *Recognizable Image*, p. 137.
44 *The Hieroglyphics of a New Speech*, pp. 164–65; *Poets of Reality*, pp. 306–07.
45 *The Visual Text of William Carlos Williams*, pp. 21 and 14, n. 7. Although I agree with Sayre's reservations about those who accept Williams's prose statements without noticing how the poems work, I maintain that Williams's prose disclaimers grow out of his attraction to a poetics that did not fully satisfy him.
46 Williams describes Sheeler's later art as "subtler particularization, the abstract if you will" (*SE* 233).
47 In a 1950 interview with John W. Gerber, transcribed by Emily Mitchell Wallace, Williams noted: "Off center, that's what eccentric means" (*SSA* 15).
48 See *The Meanings of Measure*, especially pp. 106–12, on Williams's exploration of expressive and mimetic theories of art. J. Hillis Miller, in "Williams' *Spring and All* and the Progress of Poetry," *Daedalus* 99(1970):415–29, also discusses Williams's early simultaneous attraction to theories of the imagination as mimesis, revelation, and creation *ex nihilo*. Finally, see David Walker's suggestion in *The Transparent Lyric: Reading and Meaning in the Poetry of Stevens and Williams* (Princeton: Princeton University Press, 1984), p. 23, that Williams's poems deliberately refuse to let his readers "rest secure in any fixed, comfortable pattern of seeing or thinking." In view of Cushman's suggestions about how the poems express "antagonistic cooperation" (p. 112) and in view of the double status, as metaphor and as objective description, of the concluding image of "Young Sycamore," there may well be a subdued pun on the phrase "the horns of a dilemma" in the image of the horned twigs.
49 See Cushman's arguments on stanzaic patterns and Williams's use of measure as both scheme and trope, as well as Sayre's description of how Williams's understanding of Cubism and Objectivism informs his shorter, stanzaically patterned poems that consist of apparently objective description (*The Meanings of Measure*, pp. 3, 93–99; *The Visual Text of Williams Carlos Williams*, pp. 71–73).
50 In *The Transparent Lyric*, p. 20, David Walker notes a possible source of this idea of impersonality in the writings of Cubist and Precisionist artists and further discusses how impersonality did not necessarily mean lack of emotion. My argument is not that Williams equated the impersonal with the unemotional, but that he at times had other definitions of emotion—as *loosening* and as rooted in a person—which were not as easily reconciled with the aesthetic under discussion, given the American audience for whom he wrote.

51 *Selected Prose of T. S. Eliot*, ed. Frank Kermode (New York: Farrar, Straus and Giroux, 1975), p. 40. Obviously, I have ignored the complexity of Eliot's discussion of the artist's extinction of personality in order to make my point about the common association between impersonality and science. I could equally have invoked others; for example, Auden rebukes Spender for his love of classical music by saying that "the poet's attitude must be absolutely detached, like that of a surgeon or scientist" (cited by Julian Symons, *The Thirties: A Dream Revolved* [London: Cresset Press, 1960], p. 13).

52 The image may be borrowed from Moore, who reviewed Roget's *Thesaurus*, defining it, typically, in more biological terms as "analogous to the laboratory scientist's classification of species" ("Briefer Mention," *Dial* 80[May 1926]:431); she also called Williams "our Audubon" in her speech presenting him with the Russell Loines Memorial Award in 1951 (YALC), again borrowing from other scientific disciplines.

53 Cecelia Tichi has pointed out that Williams's descriptions of poems as machines and of words cleansed in acid baths may draw on the image of words as precision parts, which the poet then creatively assembles ("Medicine and Machines Made of Words").

54 "Belly Music," *Others* 5(July 1919):26.

55 Tichi describes how Williams's training, among other things, would have made him see plants as well as human beings as machines or structures, that is, "to see the material world virtually as an engineer would" ("Medicine and Machines Made of Words").

56 Henry Sayre agrees that "Precisionism, Objectivism and the aesthetics of the machine . . . were finally antagonistic to at least a part of [Williams's] sensibility" ("After *The Descent of Winter*," p. 24).

57 "What is the Use of Poetry," p. 10.

58 A discussion of two similar impulses in Williams's work can be found in *The Visual Text of William Carlos Williams*, chapters 1 and 2. Sayre proposes that the dialectic between mind and matter, or abstraction and reality—roughly corresponding to the two aesthetics I discuss—is significantly modified in *Paterson*'s images of modern physics (pp. 111–12), although he does not discuss why physics, specifically, might serve such a function. In "Twentieth Century Limited," pp. 65–66, Tichi notes that Williams at times justified his poems of process as well as his imagistic poems with a poetics framed in technological language.

59 Letter to Jim Higgins, undated, cited in *Williams: The American Background*, p. 42.

60 For discussions of Williams's fluid style, see *Williams: An American Artist*, pp. 34–35; *The Early Poetry of William Carlos Williams*, pp. 155–56; and Reed Whittemore, *William Carlos Williams: Poet from*

Jersey (Boston: Houghton Mifflin, 1975), p. 122. For passing references to modernist analogies between a fluid style and current scientific theories, see *The Pound Era*, p. 153, and *Poet from Jersey*, pp. 115–16.

61 Fluidity is Williams's own image; for instance, in 1936 he characterized the "demonic power of the mind . . . [that gives rise to and] is the reason for the value of poetry . . . [as] a fluid speaking" (*SSA* 98). In the essay, "How to Write," Williams distinguishes between fluid speaking and actual, object-like words on the page, although he suggests that words are given value by the creative flux that informs them. In the poems that are most obviously subjective in content and loosely structured in form, however, Williams experiments with a style that tries to imitate "fluid speaking."

62 10 January 1938 letter to Constance Rourke, cited in *Williams: The American Background*, p. 62. Williams's praise of Sheeler is found in his 1939 introduction to a Museum of Modern Art catalogue on the artist and his 1954 note from *Art in America* (Williams's essays are reprinted in *A Recognizable Image*, 140–48).

63 I am here following a suggestion made by Emily Mitchell Wallace in "The Forms of the Emotions . . . the Pointed Trees," in *William Carlos Williams*, ed. Charles Angoff (Cranbury, New Jersey: Associated University Presses, 1974), p. 29. Wallace quotes his description of crystalline emotions while discussing the importance of feelings for Williams, feelings that include "Dionysian rapture." Without denying that the poem also celebrates observation and composition, as Wallace says, I want to underline the way in which the poetic style points toward, or is a trope for, such rapture, something the usual descriptions of a machine style did not encompass. See also *American Beauty*, pp. 132–33, where Tapscott discusses this same passage in terms of the "verbal energies [that] mediate between external and internal forces."

64 *Conceptions of Reality in Modern Poetry*, pp. 50–51, discusses the reference to the geometry of inner and outer reality in this poem. While I agree that Williams retains here a view of the geometrical or structural, I am suggesting that his attention shifted to images of fire and fluidity. See also *The Transparent Lyric*, p. 151, on Williams's use of a related style in "The Sea-Elephant," in which David Walker also argues that Williams's style is an evocation of the act of the mind.

65 *Williams: The American Background*, pp. 47–52, 65–66.

66 "Twentieth Century Limited," pp. 49–62.

67 Williams's letter on *Science and the Modern World* is cited in *Williams: The American Background*, p. 48. Williams was also reading *Four Lectures on Relativity and Space*.

68 All references will be to the early version of the poem found in *Contact*,

no. 4 (Summer 1921):2–4. The poem, but not the preface to it, is reprinted in *Myth and Muse*, pp. 195–98. Donley and Friedman note that Williams's poem may well mark Einstein's first appearance as a muse; my reading of this poem, and of Williams's view of Einstein generally, differs from Donley and Freidman's (pp. 68–71) in that I see Williams as having some ambivalence about Einstein's popular success. Further, although I do agree that Williams's use of relativity led him to the mature work of *Paterson* (pp. 131–33), I take issue with the characterization of Williams's earlier work as "the stasis of Imagism" (p. 71).

69 See *No Place of Grace*, pp. 151–53.

70 See *Myth and Muse*, pp. 17–18, on Einstein's popular image.

71 "Prof. Einstein Here, Explains Relativity; 'Poet in Science' Says It Is a Theory of Space and Time, But It Baffles Reporters," *New York Times*, 3 April 1921, p. 1; cited in "A little touch of / Einstein in the night—," p. 11.

72 "Reply to A Young Scientist Berating Me Because of My Devotion to a Matter of Words," p. 28. Note Williams's pun on "matter."

73 In "Reply to A Young Scientist," in fact, Williams calls scientists "poor fishes" (p. 28), recalling the poisonous fish of "St. Francis Einstein."

74 See "Relativity and Radioactivity in William Carlos Williams' *Paterson*," and *The Meanings of Measure*, p. 113.

75 *The Visual Text of William Carlos Williams*, p. 60. *Myth and Muse* traces the actual connection between Einstein's discoveries and the production of the bomb as opposed to the popular understanding of that connection (pp. 171–76).

76 This is to qualify, but not to deny, Emily Mitchell Wallace's suggestion in "The Satyrs' Abstract and Brief Chronicle of Our Time," p. 151, that Williams proposes the imagination can overcome the tyranny of modern images.

77 "Roads to Einstein," p. 174.

78 Hartley's letter to Stieglitz of June 1915 is cited in *Williams: The American Background*, p. 42, as an important influence on Williams's experiments in *Kora in Hell*.

79 Thus the "Primavera" group, including "The Sea-Elephant," is neither a throwback to early experiments nor unrelated to the poems more immediately preceding it, as *The Early Poetry of William Carlos Williams*, p. 156, suggests. Rather, the poems' fluid style is a way of emphasizing both motion and attention to the particular and of incorporating some of the features found in poems like "Young Sycamore." See also the discussion of "The Sea-Elephant" as various voices in *The Transparent Lyric*, pp. 151–56.

80 See *The Visual Text of William Carlos Williams*, pp. 49 and 72.

81 See *The Hieroglyphics of a New Speech*, p. 173.
82 "Medicine and Machines Made of Words"; "Towards the Crystal," pp. 197–98.
83 *Axel's Castle*, pp. 157–58. Wilson, of course, draws on Whitehead (see *Axel's Castle*, p. 3). *Myth and Man*, pp. 140–48, discusses how quantum theory, in particular, posits a blend of subject and object in the process of observation parallel to the blend suggested by the style of many modern writers, although Donley and Friedman primarily discuss writers of fiction in this context.
84 "The Name and Nature of Modernism," pp. 44 and 48. Bradbury and McFarlane focus on the tensions between Classicism and Romanticism, in particular.
85 See *A New World Naked*, p. 325, for a discussion of Williams's defense of literature during the Depression. Family pressures made this period a time of personal depression for Williams as well.
86 "The Sense of the Past," p. 29. Trilling is discussing literary history generally in this essay, which does not mention Williams.

CHAPTER 5

1 *A Homemade World*, p. xiv.
2 See also Robert Beloof, "Prosody and Tone: The 'Mathematics' of Marianne Moore," in *Marianne Moore: A Collection of Critical Essays*, ed. Charles Tomlinson (Englewood Cliffs, New Jersey: Prentice-Hall, 1969), pp. 144–49.
3 From an interview with Howard Nemerov, *Poetry and Criticism* (Cambridge, Massachusetts: Adams House and Lowell House Printers, April, 1965), n.p.
4 "Concerning the Marvelous," Rosenbach, TS [typescript], reviewing a show at the Museum of Modern Art, 9 January 1937, p. 1.
5 Emmy Veronica Sanders, "America Invades Europe," *Broom* 1(November 1921):89. Work by Moore appeared in the January 1922 issue of *Broom*. Moreover, in her interview with Donald Hall, she mentions being aware of *Broom* from its inception (*R* 269), making it quite possible that she read Sanders's article. It should be added that Sanders linked the European Dadaists with the love of technology, and wanted Americans to develop their own aesthetic (see *Skyscraper Primitives*, pp. 118–21, on the changing editorial position of *Broom* concerning technology).
6 See "Made in America," *Broom* 2(June 1922):270 and "After and Beyond Dada," *Broom* 2(July 1922):347.
7 See, for instance, *Hound & Horn* 1(December 1927):157.
8 Moore quotes Eliot's remark on James, which appeared in the August

1918 issue of the *Little Review*, edited by Pound, where Eliot says James waved aside all "this show commercialism wh[ich] Americans like to present to the foreign eye" (Reading diary for 1924–1930, Rosenbach 1250/5, MS, p. 68). Moore's work appeared in *Others*, the *Egoist*, and *Poetry* in 1915; in *Broom* by 1922; and in *Hound & Horn* by 1932.

9 "America Invades Europe," p. 92.

10 Reading diary for 1921–1922, Rosenbach 1250/3, MS, p. 61; Moore's notes indicate she was reading *The Letters of Henry James* at this time.

11 "America and the Young Intellectual," unpublished review, circa 1919–1920, Rosenbach, TS. See chapter 1 on Santayana's review of Stearns's book, which appeared in the *Dial* where Moore would surely have read it.

12 Moore was an active contributor to the *Dial* well before the official announcement of her editorial status in June 1926. Nicholas Joost and Alvin Sullivan, *The 'Dial,' Two Author Indexes: Anonymous & Pseudonymous Contributors; Contributors in Clipsheets*, Bibliographical Contributions, no. 6 (Carbondale, Illinois: Southern Illinois University Libraries, 1971), make clear which editorial pieces were written by Moore.

13 "Comment," *Dial* 79(October 1925):355; Keyserling's writing appeared in the same issue of the *Dial*. *Dial* 80(March 1926):265; Moore also quotes from *The American Scholar* (by Emerson): "We, it seems, are critical. . . . We cannot enjoy anything for hankering to know whereof the pleasure consists; we are lined with eyes; we see with our feet" (p. 266).

14 "Comment," *Dial* 81(October 1926):357–58. A note on Čapek's article from the *New York Times*, 16 May 1926, appears in Moore's reading diary for 1924–1930, Rosenbach 1250/5, MS, p. 125, although the public comment on Čapek does not appear until October. For further discussion of the common association between America and speed, see "Twentieth Century Limited," pp. 50–54. While Moore's comment is on technological speed and American restlessness, it is worth noting that Čapek's 1924 novel, *Krakatit*, translated in 1925, was about the new physics as well as technology. The novel is discussed in *Myth and Muse*, pp. 83–84.

15 The quotation is from a letter written to her brother (probably in January 1916) after an unsuccessful attempt to find work at a Philadelphia newspaper. The statement is cited in *The Poet's Advance*, p. 52.

16 *Dial* 82(March 1927):267.

17 See Marie Borroff, *Language and the Poet: Verbal Artistry in Frost, Stevens, and Moore* (Chicago: University of Chicago Press, 1979), pp. 80–89, 102, 132, on Moore and advertising. See also Moore's comment

that we "read advertisements as we read the body of a magazine and like the impression of energy and ability" in *Dial* 83(November 1927):450.

18 Moore's notes on an article about the invention of the typewriter provide another link between typewriters and sewing machines, again alluding to the look of technological inventions. Both are called "beautiful" (Reading diary for 1923, Rosenbach 1250/4, MS, p. 26). Many entries in the diaries testify to Moore's interest in industrial design.

19 Reading diary for 1924–1930, Rosenbach 1250/5, MS, p. 117. See also Moore's attention to sound in "Humility, Concentration, Gusto" (*Pred* 12–20).

20 In "Briefer Mention," *Dial* 78(January 1925):75, Moore identifies pliant dexterity as characteristic of a modern style. Note that Williams's associations with speed and rapidity were somewhat different. As he wrote to Robert Carlton Brown, "[when] I begin to speed up . . . the sense of *power* that comes with that is the best tonic I know" (*SL* 114, emphasis added).

21 Moore also links science with precision and with detail several times in her "Interview with Donald Hall" (*R* 255 and 273). It is worth noting that Moore's definitions of accuracy and precision, unlike Williams's, tend to include tidiness. This may not be surprising given Moore's experiences as a librarian and an editor in contrast with Williams's career as an overworked doctor and on-the-spot diagnostician.

22 Moore said sympathetically that "Wallace Stevens was really very much annoyed at being . . . compelled to be scientific about what he was doing," in her "Interview with Donald Hall" (*R* 273).

23 All quotations are from the 1932 version of the poem as it first appeared in *Poetry Magazine*, unless otherwise noted. "The Student" was first published as the middle section of "Part of A Novel, Part of A Poem, Part of A Play" (*Poetry* 40[June 1932]:122–26). It was reprinted in *Furioso* 1(Summer 1941):22–23, and reappeared on its own in the 1941 volume, *What Are Years*, with some revisions.

24 "Comment," *Dial* 80(June 1926):532. Notes on Roger Fry's *Art and Commerce* appear in the later reading diary for 1930–1943, Rosenbach 1250/6, MS, p. 110.

25 "Comment," *Dial* 79(August 1925):177.

26 Reading diary for 1930–1943, Rosenbach 1250/6, MS, p. 42, taken from H. McBride's article of 12 December 1931 in the *New York Sun*. Dr. Valentiner was the curator of decorative arts at the Metropolitan Museum of Art in New York from 1908–1914, after which he moved to the Detroit Art Institute, where he became director in 1924. See Wallace Stevens's attention to Valentiner (*L* 300).

27 *New Statesman*, 29 July 1916, p. 405; Alan Holder, *A Study of Henry*

James, Ezra Pound, and T. S. Eliot: Three Voyagers in Search of Europe (Philadelphia: University of Pennsylvania Press, 1966), p. 60, points out that the review, which was published anonymously, is by Eliot. Moore's diary also includes notes on how America's inventors were not primarily interested in money: there is a passage from *Transcript*, 20 June 1914, on how Edison was "unbusiness-like" (reading diary for 1907–1915, Rosenbach 1250/1, MS, p. 88) and from G. S. Lee, *Survey*, 7 February 1914, on what a shame it was that Bell had to stop inventing and start marketing his inventions (ibid., p. 95). See also Moore's attention to Pound's characterization of America as "*Midas lacking a Pan*" (*R* 153) and to Pound's dislike of "America's imperviousness to culture" (*R* 160).

28 H. D., "Marianne Moore," *The Egoist* 3(August 1916):119.

29 "Comment," *Dial* 79(August 1925):177.

30 Reading diary for 1907–1915, Rosenbach 1250/1, MS, p. 72.

31 See, for instance, "Is Nationalism Moribund?" p. 682. That Moore embraced at least her version of science more wholeheartedly than many of her peers embraced their versions is evidenced also in her distaste for the vocabulary found in Donald Allen's anthology, *The New American Poetry*, which she contrasts with "the vocabularies of science, which are creative" (*R* 242).

32 Reading diary for 1916–1921, Rosenbach 1250/2, MS, p. 91.

33 See Steinman, "Moore, Emerson and Kreymborg: The Use of Lists in 'The Monkeys'," *Marianne Moore Newsletter* 4(Spring 1980):7–10. See also Moore's comment in "A Grave," where she comes to terms with the inhuman sea by populating it with human markers such as lighthouses and bell-buoys: "it is human nature to stand in the middle of a thing" (*Comp* 49).

34 "Preface," *Democracy and Education*, p. v; Reading diary for 1916–1921, Rosenbach 1250/2, MS, p. 40, mentions *Democracy and America*.

35 Reading diary for 1921–1922, Rosenbach 1250/3, MS, p. 43, on a review by Nelson Antrim Crawford. Crawford's article, "Unity Made Vital," *Poetry* 18(September 1921):338, also notes that science and art "may be popularly regarded as enemies."

36 Moore says in a *Dial* piece on completeness versus miscellany: "However expressive the content of an anthology, one notes that a yet more distinct unity is afforded in the unintentional portrait given, of the mind which brought the assembled integers together" ("Comment," *Dial* 82[May 1927]:450). As an early diary note on a December 1913 article about censorship suggests, for Moore the "moral quality of a work of art . . . [lies] in its relation to the percipient mind" (Reading diary for 1907–1915, Rosenbach 1250/1, MS, p. 62).

37 In the announcement of the presentation of the *Dial* Award to Burke in *Dial* 86(January 1929):90. As Moore admits in "Idiosyncrasy and Technique," "it is a curiosity of literature how often what one says of another seems descriptive of one's self" (*R* 178). See also, on the subject of the relationship between a poetic style and the age, her restrained criticism of Charles Cotton's poems: "An age of brilliance ought not to be commemorated in four hundred and twenty pages of conventional love, bucolic conviviality, and elaborate idleness" ("Briefer Mention," *Dial* 76[March 1924]:289).

38 "Comment," *Dial* 83(October 1927):358.

39 See Bonnie Costello, *Marianne Moore: Imaginary Possessions* (Cambridge: Harvard University Press, 1981), on Moore as a poet of process and, in particular, on how many of Moore's apparent descriptions of real objects borrow from verbal accounts of art objects. I am indebted to Costello's reading of Moore throughout this chapter. Costello discusses "The Student" on pages 248–49. As Costello makes clear, the view of Moore as a poet of limited energy, who concentrates upon descriptions of small objects in the external world, misses the mark; but see Bernard F. Engel, *Marianne Moore*, Twayne's United States Authors Series, no. 54 (New York: Twayne Publishers, 1964), p. 159, on Moore's emphasis on "the thing itself," or Randall Jarrell, "The Humble Animal" and "Her Shield," in *Poetry and the Age* (New York: Alfred A. Knopf, 1953), pp. 179–207.

40 "Sample Prose Piece / The Three Letters," *Contact* 1(Summer 1921):10–11.

41 See Ann L. Hayes, "On Reading Marianne Moore," in *A Modern Miscellany*, Carnegie Series in English, no. 11 (Pittsburgh: Carnegie–Mellon University, 1970), pp. 7–11, for a discussion of Moore's craftsmanship in this poem.

42 See *Imaginary Possessions*, p. 107: "We understand Moore's poems best if we consider the movement of their composition, rather than the theme or the representational subject."

43 For a discussion of Bourne's stance in 1916, see *The End of American Innocence*, pp. 326–28. Moore's entry, from her reading diary 1916–1921, Rosenbach 1250/2, MS, p. 28, seems to predate the publication of Bourne's best known statements about education in "Trans-National America," *Atlantic Monthly* 118(July 1916):86–97, of which Moore may have been reminded when she wrote in Bourne's name.

44 Bourne's question is found in "Trans-National America," p. 86, where he describes the failure of the American melting pot. Bourne also discusses the pioneer mentality concerned with "material resources" (p. 87), and speaks of the need for a distinctly American education and

literature, along the lines of Whitman, Emerson, and James (pp. 88–93). The article concludes, comparing America with France, that "the contribution of America will be an intellectual internationalism which goes far beyond the mere exchange of scientific ideas and discoveries and the cold recording of facts" (p. 94).

45 Reading diary for 1916–1921, Rosenbach 1250/2, MS, pp. 39–40. In the 1916 *Democracy and Education* mentioned by Moore, Dewey's argument was that the purpose of education was primarily "to keep alive a creative and constructive attitude" and to liberate human intelligence and human sympathy (*Democracy and Education*, p. 231).

46 Reading diary for 1916–1921, Rosenbach 1250/2, MS, p. 121.

47 Reading diary for 1916–1921, Rosenbach 1250/2, MS, p. 122.

48 Reading diary for 1921–1922, Rosenbach 1250/3, MS, p. 55.

49 Reading diary for 1930–1943, Rosenbach 1250/6, MS, pp. 162–72, for instance, mentions such titles as *The World of Science*, *Pathways in Science*, *Scientific Monthly*, and *Francis and Roger Bacon and Modern Science* from the Science Press.

50 Reading diary for 1930–1943, Rosenbach 1250/6, MS, p. 172, from Dr. Fernando Sanford's *Francis and Roger Bacon and Modern Science*.

51 Reading diary for 1916–1921, Rosenbach 1250/2, MS, p. 142, quoting from *The Medieval Mind*, vol. II, p. 516.

52 Bonnie Costello, "'To a Snail': A Lesson in Compression," *Marianne Moore Newsletter* 3(Fall 1979):14.

53 Moore's journals include notes on Burbank (reading diary for 1921–1922, Rosenbach 1250/3, MS, p. 84 and reading diary for 1924–1930, Rosenbach 1250/5, MS, p. 88) and on Darwin (reading diary for 1916–1921, Rosenbach 1250/2, MS, p. 127). See also "Comment," *Dial* 79(November 1925):443, in which Burbank is praised for being "an observer."

54 Reading diary for 1916–1921, Rosenbach 1250/2, MS, p. 127; notes on the Book of Isaiah from the first volume of the *Expositor's Bible*.

55 John Dewey, *Reconstruction in Philosophy* (New York: Henry Holt, 1920), p. 127.

56 Rosenbach notes, pp. 6–7. In "Aids to Precision," Rosenbach TS, p. 10, a piece related to "Feeling and Precision" (*Pred* 3–11), Moore recommends reading as an aid to writing, saying, "In my own case I am inclined to put first, *scientific works* such as Darwin's."

57 See "Comment," *Marianne Moore Newsletter* 3(Fall 1979):4; "Comment," p. 3, and "Observations," p. 5, in *Marianne Moore Newsletter* 5(Spring 1981).

58 "Briefer Mention," *Dial* 80(May 1926): 431.

59 Reading diary for 1916–1921, Rosenbach 1250/2, MS, p. 36.

60 Reading diary for 1916–1921, Rosenbach 1250/2, MS, p. 66.

61 It may be relevant that "The Student" was written not too many years after the 1925 Scopes trial.

62 Reading diary for 1916–1921, Rosenbach 1250/2, MS, p. 66.

63 Reading diary for 1916–1921, Rosenbach 1250/2, MS, p. 68; I have silently added some punctuation and provided correct spelling. On the same page of her reading diary, just after this impassioned plea for accuracy of information, Moore added notes from Burke on the difficulty of shearing wolves, quoting lines that appear in the later version of "The Student," and reinforcing the suggestion that these speculations on science and poetry inform the poem.

64 See *Imaginary Possessions*, p. 44, on how Moore typically creates the illusion of standing on both sides of an argument.

65 Reading diary for 1916–1921, Rosenbach 1250/2, MS, p. 122.

66 See *The Pragmatic Movement in American Philosophy*, pp. 59–71, for an account of the American pragmatists' problems with similar issues.

67 Lewis Mumford, "The City," in *Civilization in the United States*, p. 9.

68 *Marianne Moore*, p. 45; Jarrell, "On Being Modern With Distinction," in *Marianne Moore: A Collection of Critical Essays*, p. 102. See also *Poetry and the Age*, p. 189, where Jarrell finds that the end of "New York" makes "the best and truest case that can be made out for Americans." Finally, see Moore's qualified assent to Claude Bragdon's comment on the skyscraper as a symbol of "ruthless . . . aggression," "Comment," *Dial* 86(April 1929):359.

69 See Moore's remark on how Henry James, a "characteristic American," was for freedom even from "an excellent cause" (*R* 136).

70 Moore may, indeed, have been thinking of the Americans who were living in France at the time, including Bourne to whom the general comparison between French and American education may be indebted whether or not the last line specifically refers to him.

71 Moore's reading diary for 1907–1915, Rosenbach 1250/1, MS, p. 116, includes notes from an article ["The Stout Lady Buys a Dancer," pp. 301–02] by Horace Holley in *New Republic*, 24 April 1915, on how the love of art requires self-consciousness.

72 T. S. Eliot, review of *Poems* and *Marriage*, *Dial* 75(December 1923): 597.

73 See "William Carlos Williams and the Efficient Movement," and *Anti-Intellectualism in American Life*, p. 285. Yet also see "Lewis Mumford and The Myth of the Machine," p. 22, on the early belief that some aspects of modernity, including Taylorism, might work given a differently structured workplace.

74 For an account of the labor history of Paterson, "a Wild West outpost of

industrialism," see Christopher Norwood, *About Paterson: The Making and Unmaking of an American City* (New York: Harper and Row, 1975), especially p. 41.

75 See, for example, "Comment," *Dial* 83(October 1927):360.

76 See Moore's self-consciousness about her audience's problems with her work in, for example, "Interview with Donald Hall" (*R* 259).

CHAPTER 6

1 "Enquiry," *Twentieth Century Verse*, American Number, 12–13(October 1938):114.

2 See Melita Schaum, "Concepts of Irony in Wallace Stevens' Early Critics," *Wallace Stevens Journal* 9(Fall 1985):85–97, for a review of debates about Stevens as a writer in America.

3 "Enquiry," p. 112.

4 See Glen MacLeod, *Wallace Stevens and Company: The Harmonium Years, 1913–1923* (Ann Arbor: UMI Research Press, 1983); the essays in *Wallace Stevens: A Celebration*, ed. Frank Doggett and Robert Buttel (Princeton: Princeton University Press, 1980); Peter Brazeau, *Parts Of A World: Wallace Stevens Remembered, An Oral Biography* (New York: Random House, 1983), and Milton J. Bates, *Wallace Stevens: A Mythology of Self* (Berkeley and Los Angeles: University of California Press, 1985). Joan Richardson's biography is the most complete source of information about Stevens to date. The first volume, *Wallace Stevens, A Biography: The Early Years, 1879–1923* (New York: Beech Tree Books, William Morrow, 1986), unfortunately appeared too late for me to make full use of it here, although throughout this chapter I am indebted to Prof. Richardson's conversations with me on Stevens.

5 See *The Harmonium Years*, and Louis L. Martz "'From the Journal of Crispin': An Early Version of 'The Comedian As the Letter C'" in *Wallace Stevens: A Celebration*, p. 10.

6 *The Harmonium Years*, pp. 33–40.

7 6–7 February 1909, letter to Elsie Viola (Moll) Stevens, WAS 1815, Huntington Library.

8 9 May 1909, letter to Elsie Viola (Moll) Stevens, WAS 1843, Huntington Library.

9 See also *Letters*, pp. 145, 166, 168; *Souvenirs and Prophecies*, p. 96, and *A Mythology of Self*, pp. 44 and 61.

10 See *Letters*, p. 50 or *Souvenirs and Prophecies*, pp. 224–25, for instance. This is not to quarrel with Peter Brazeau's discussion of Stevens's affection for New York shops, galleries, restaurants, and friends in the 1930s and 1940s, after he had moved to Hartford. See Peter Brazeau, "A Trip

in a Balloon: A Sketch of Stevens' Later Years in New York," in *Wallace Stevens: A Celebration*, especially pp. 342–43.

11 See *A Mythology of Self*, p. 87. *Wallace Stevens, A Biography, A Mythology of Self*, chapters 1 and 2, and Richard Ellmann, "How Stevens Saw Himself," in *Wallace Stevens: A Celebration*, pp. 154–58, give a fuller picture of Stevens's relationship with his father. Stevens's father also identified the need for a profession as a specifically American preoccupation (*SP* 71).

12 See *A Mythology of Self*, p. 30, and *The Harmonium Years*, pp. 10–12, on Stevens's relationship to the aesthetes and poetry of the 1890s. Stevens offers perhaps his clearest rejection of aestheticism, assigning it to the "diplomats of the cafés," in "Owl's Clover" (*OP* 57). His firm stance is related to the charges of dandyism leveled against his own early poetry, and his need to explain how it was not effeminate or trivial.

13 Samuel French Morse, *Wallace Stevens: Poetry as Life* (New York: Pegasus, 1970), p. 127, attributes this phrase to Santayana's *Poetry and Religion*. Stevens later returned to this question of the uses of beauty after reading Paul Elmer More (*SP* 212).

14 For a record of books known to have been in Stevens's library, see J. M. Edelstein, "The Poet as Reader: Wallace Stevens and His Books," *Book Collector* 23(1974):53–68; Peter Brazeau, "Wallace Stevens at the University of Massachusetts: Check List of An Archive," *Wallace Stevens Journal* 2(Spring 1978):50–54; and Milton J. Bates, "Stevens' Books at the Huntington: An Annotated Checklist," *Wallace Stevens Journal* 2(Fall 1978):45–61, and 3(Spring 1979):15–33. Mauron's book is in the Huntington Library Collection.

15 Charles Mauron, *Aesthetics and Psychology* (London: Hogarth Press, 1935), p. 105; see also 29, 57–58.

16 See *Aesthetics and Psychology*, pp. 31–38, 70, and *Souvenirs and Prophecies*, pp. 38–39.

17 Moore is quoted from worksheets for "The Plumet Basilisk," which are reprinted in Bonnie Costello, "Marianne Moore's Debt and Tribute to Wallace Stevens," *Concerning Poetry* 15(Spring 1982):29; Stevens's remark is from "An Enquiry," p. 15.

18 Yet Stevens was still writing sonnets in 1906 (*L* 92); in 1946, commenting on the "realism" of the situation in England, he echoed his early judgment on sonnets saying that American poetry, his own in particular, might sound "terribly out of place" (*L* 524) in postwar England.

19 Cited in Holly Stevens, "Flux," *Southern Review*, Wallace Stevens Centennial Issue, 15(Autumn 1979):773.

20 *Grammar of Motives*, pp. 224–26, describes Stevens's theory of poetry as a form of scientism. Burke implies that Stevens was unaware of his

affiliations with science, which, by the 1940s, was not entirely true. Still, Burke's argument on the scientism implicit in certain forms of idealism suggests part of the reason that Stevens ultimately could amend some of his early ideas about the antagonism between poetry and science.

21 *Science and the Modern World*, p. 3. Alan Perlis, *Work in Progress*, Chapter 3 ("Poetry and Physics: A Confrontation Across the Spheres"), pp. 86–94, nicely outlines three often contradictory commonplaces about the relationship between poetry and science: poetry deals with the timeless, science with materialistic and fleeting particulars in time; poetry is concerned with the concrete, science stays with the general; and science uses reason while poetry requires imagination.

22 *Aesthetics and Psychology*, pp. 90–91; see also pp. 40–48.

23 The Huntington Library has the original manuscript; the piece was originally published in *Opus Posthumous*, having been mistaken for Stevens's own since it was written in his hand (see *OP* vii).

24 Much of the following discussion draws heavily on *Poets of Reality*, pp. 1–12, 217–84.

25 See Harold Bloom, *Wallace Stevens: The Poems of Our Climate* (Ithaca: Cornell University Press, 1977), p. 46.

26 See Edward Kessler, *Images of Wallace Stevens* (New Brunswick: Rutgers University Press, 1972), pp. 112–13, and *Poems of Our Climate*, pp. 45–46.

27 Martha Strom, "Wallace Stevens' Revisions of Crispin's Journal: A Reaction Against the 'Local'," *American Literature* 54(May 1982):265–68, 276. Milton Bates discusses Stevens's naturalism, but he expressly says Stevens's style is naturalistic primarily in avoiding didacticism and romanticism (*A Mythology of Self*, pp. 132–34).

28 See " 'From the Journal of Crispin'," especially pp. 7–9 for passages edited out of "The Comedian As the Letter C," which show Stevens's original attempt to include the local. See also "Wallace Stevens' Revisions of Crispin's Journal," pp. 258–76.

29 Christopher P. Wilson, *The Labor of Words: Literary Professionalism in the Progressive Era* (Athens: The University of Georgia Press, 1985), pp. 198–200.

30 *The Labor of Words*, pp. 20, 28, 34–39.

31 See *The Necessary Angel*, p. 20–21. Stevens did see daily newspapers, like airplanes, as offering the excitement of the new, but he also discussed how quickly such novelty faded (*L* 145; *SP* 239).

32 See "Wallace Stevens' Revisions of Crispin's Journal," pp. 265–66, which distinguishes between Williams's use of the local and the strict idea of local detail found, for example, in James Oppenheim's "Poetry—Our First National Art," discussed in chapter 1 above.

33 The *New Republic,* 15 July 1936, p. 305.

34 For a description of the politics of those associated with the various journals discussed here, and the shared concern over the role of the arts in America, see Richard H. Pells, *Radical Visions and American Dreams: Culture and Social Thought in the Depression Years* (New York: Harper and Row, 1973), especially pp. 151–93.

35 "Turmoil in the Middle Ground," p. 42.

36 See Joseph N. Riddel's argument in *The Clairvoyant Eye: The Poetry and Poetics of Wallace Stevens* (Baton Rouge: Louisiana State University Press, 1965), p. 126, that these lines constitute Stevens's defense of *Harmonium.* It is true that Stevens later calls his central poet the "spokesman at our bluntest barriers" (*CP* 397) and yet he adds there that such a poet may evade us (*CP* 396) so that unqualified bluntness is not necessarily a sign of Stevens's voice of approval. As he has Ozymandias say to the naked Nanzia Nunzio, "the bride / Is never naked. A fictive covering / Weaves always glistening from the heart and mind" (*CP* 396).

37 See, for instance, "Sunday Morning," "The Sense of the Sleight-of-Hand Man," or "Ploughing on Sunday." James Baird, in *The Dome and The Rock: Structure in the Poetry of Wallace Stevens* (Baltimore: Johns Hopkins Press, 1968), p. 51, discusses Stevens's usual references to Sundays.

38 Ruth Lechlitner, "Imagination as Reality," *The New York Herald Tribune Books,* 6 December 1936, p. 40.

39 Stevens, who spent time in the Botanical Gardens, was probably aware that owl's clover is a false clover. In yet another sense, then, the poem betrays a recognition that the poetry of the day's news might not be authentic poetry.

40 See *Wallace Stevens: The Making of the Poem,* p. 120. Peter Brazeau, in *Parts Of A World,* p. 245, suggests that Stevens's silence was also health related.

41 Cited in *Parts Of A World,* p. 244.

42 Not only had his family suggested that literature and making a living might be incompatible, but in Stevens's library holdings at the Huntington library, in Herbert W. Paul's *Matthew Arnold* (New York: Macmillan, 1903), p. 176, Stevens (who inscribed the book in 1918) has underlined a description of how having to work for a living "almost dried up the poetic vein" in Arnold.

43 See *A Mythology of Self,* p. 162, on Stevens's financial well-being during the Depression.

44 See also *A Mythology of Self,* p. 190, on how contemporary concerns inform "The Man With The Blue Guitar."

45 Stevens seems to have felt at times that law in particular (the career on

which he embarked) was at odds with imaginative work; a 1908 letter to Elsie Moll explains that "law is mostly thinking without much result," an activity Stevens implicitly contrasts with more joyful poetic thinking without an eye to results when he says he escaped office pressures by trying to play his guitar (although the guitar playing here is literal) (*SP* 201).

46 See Charles Berger, *Forms of Farewell: The Late Poetry of Wallace Stevens* (Madison: University of Wisconsin Press, 1985), pp. 30–38, 82–84, on Stevens's response to the war and to the Depression in his longer poems written after "The Man With The Blue Guitar."

47 See *Letters*, p. 313, where Stevens links "The Irrational Element in Poetry" with "Owl's Clover." *Wallace Stevens: The Making of The Poem*, pp. 4–5, notes Stevens's fascination with the irrational in the 1930s.

48 Stevens's copy of I. A. Richards, *Coleridge: On Imagination; A Study of the Critical Theory of Coleridge* (London: Kegan Paul, Trench, Trubner, 1934), p. 157, contains a penciled line in the margin next to this passage. The volume is housed in the Huntington Library. Stevens's view may also be related to the emphasis on behaviorism that arose in the 1920, as noted by Johan Huizinga, *America: A Dutch Historian's Vision, from Afar and Near* (cited in "Cultural Revisions in the Twenties," p. 58).

49 It has been suggested that Stevens drew on Freud for his figure of Ananke, identifying the figure with external reality (see "Wallace Stevens at the University of Massachusetts," p. 50). See also "Sur Plusieurs Beaux Sujects," I, Huntington, MS, p. 8, on Ananke.

50 By 1944, Stevens (*L* 384) wrote that he was impressed with Leo Spitzer's "History of Ideas Versus Reading of Poetry," *Southern Review* 6(Winter 1941):584–609, in which Spitzer distinguishes between poetry itself and the "biology of the genius" (p. 588).

51 In *A Mythology of Self*, pp. 124–25, Milton Bates comments on Stevens's image as part of his continuing negative response to localism.

52 Burnshaw, a poet in his own right, was far more sympathetic to Stevens than Stevens admits, and he was clearly as concerned as Stevens about the status of art in the world of the 1930s. Burnshaw later wrote that it "requires no expertness in Freud to perceive that the present reviewer's concern with Stevens' confusion was at least in part a projection of his own," in "Wallace Stevens and the Statue," *Sewanee Review* 69(Summer 1961):362.

53 Stevens's need to convince his critics was obviously in part related to his feeling that "a poet needs above everything else . . . acceptance" (*L* 433; see *L* 436).

54 The title of this section, "Sombre Figuration," refers specifically to the figure of the portent. Although the portent is similar to the giant pres-

ences that appear frequently, especially in Stevens's later poems, it is unique in that it represents the self-image not only of a generation, but of a civilization and in that it is a portent of destruction. Yet by recognizing that the portent is an "image of [the subman's] making" (*OP* 69), the poem emphasizes man's ability to reimagine, and thus redeem, history.

55 See Joseph Evans Slate, "From The Front Page: A Note on Williams' 'The Death of See'," *William Carlos Williams Newsletter* 3 (Spring 1977):16–18.

56 See A. Walton Litz, "Particles of Order: The Unpublished *Adagia*," in *Wallace Stevens: A Celebration*, p. 59, for a note on how Stevens "won through to a tenable aesthetic" in the 1940s. See also *Wallace Stevens: The Making of the Poem*, which acknowledges Stevens's changing ideas on the nature of poetry and the increasingly explicit poetics evolved in the 1940s; Doggett further discusses Stevens's concern with the normal or central and his decreased emphasis on the irrational in the late 1930s and 1940s (pp. 106, 109, 123).

57 Cited in *Parts Of A World*, p. 162.

58 *Parts Of A World*, p. 61. *Myth and Muse*, p. 84, suggests that Stevens's 1917 "Thirteen Ways of Looking at a Blackbird" already, perhaps coincidentally, draws on Einstein's theories. There is no evidence that Stevens knew of the new physics in 1917; indeed, I argue that he was drawing on the work of modern painters but did not directly see how he could relate physics to his characteristic style of writing until the 1940s.

59 Philip Wheelwright [and others], *The Language of Poetry*, ed. Allen Tate (Princeton: Princeton University Press, 1942), p. viii.

60 *Let the People Think*, pp. 97 and 49.

61 Stevens explicitly says he is using the word fancy in a Coleridgean sense; although he owned several books on and by Coleridge, in this context he may well have been thinking of Richards's 1934 book on Coleridge. The quotation used at the beginning of "The Noble Rider"—on Plato's "dear, gorgeous nonsense" (*NA* 3)—shows that Stevens had been reading Richards's book. The quotation is from *On Imagination*, p. 149; in Stevens's copy (in the Huntington) the phrase has been copied onto the flyleaf of the book.

62 From "Henri Bergson," p. 34.

63 See "Particles of Order," pp. 57–59, 62–63, 341; Beverly Coyle, "An Anchorage of Thought: Defining the Role of Aphorism in Wallace Stevens' Poetry," *PMLA* 91(March 1976):206–22, and Lisa Steinman, "A Dithering of Presences: Style and Language in Stevens' Essays," *Contemporary Literature* 21(Winter 1980):100–17, for more detailed views of Stevens's use of aphorism.

64 See Isabel G. MacCaffrey, "The Ways of Truth in 'Le Monocle de Mon

Oncle'," in *Wallace Stevens: A Celebration*, p. 216, on Stevens's use of indefinite pronouns.

65 The idea was not unusual. See also, for instance, *Physics and Philosophy*, p. 144. Stevens, however, almost certainly drew on more popular sources (and letters from friends such as Paulhan). Moreover, scientists did not necessarily sanction the interpretations given their words by literary figures.

66 Whitehead's 1925 book does not explicitly inform Stevens's writing before the 1940s, although he could have read it earlier and, in "A Collect of Philosophy," he draws on Samuel Alexander's 1920 *Space, Time, and Deity*, which Whitehead recommends in his preface (*Science and The Modern World*, p. xi). Stevens more likely encountered Alexander's book through mention of it by C. E. M. Joad, in "Another Great Victorian," pp. 280 and 282, as noted in "Sur Plusieurs Beaux Sujects," II, Huntington, MS, pp. 1–2. Some contemporary responses to Stevens's use of mostly popular sources for his ideas about philosophy as well as science are found in *Parts Of A World*, pp. 210–14.

67 *Science and the Modern World*, pp. 276 and 282. Whitehead also discusses James, whom Stevens said he admired, noting how James agrees with the new physicists in finding "the only endurances are structures of activity" (pp. 152 and 199).

68 *Science and the Modern World*, pp. 24–25, 51. See also Whitehead's suggestion that the "reality is the process . . . The realities of nature are . . . events" (p. 102).

69 See "Particles of Order," p. 75, and *Opus Posthumous*, p. 174.

70 See Peter Brazeau, " 'A Collect of Philosophy': The Difficulty of Finding What Would Suffice," in *Wallace Stevens: A Celebration*, pp. 48–49, for a discussion of Stevens's encounter with George's piece on Planck. I am grateful to Alan Perlis, whose *Work in Progress*, pp. 104–16, first provided me with a detailed discussion of Stevens's affinities with Planck and with the new physics generally.

71 *The Conceptual Development of Quantum Mechanics*, p. 329. I am grateful to Dr. Robert Reynolds and to Dr. Richard Crandall of the Reed College Department of Physics for discussing with me and instructing me in modern physics.

72 This comparison owes at least something to Stevens's earlier desire (most obvious in "The Figure of the Youth as Virile Poet") to persuade Henry Church to establish a chair of poetry, the argument having been that chairs of philosophy were more common and yet less worthy (*Parts Of A World*, pp. 183–84).

73 Indeed, the man-man is a figure rejected in turn: "Each false thing ends" (*CP* 280).

Index